STRANGE
ENCOUNTERS

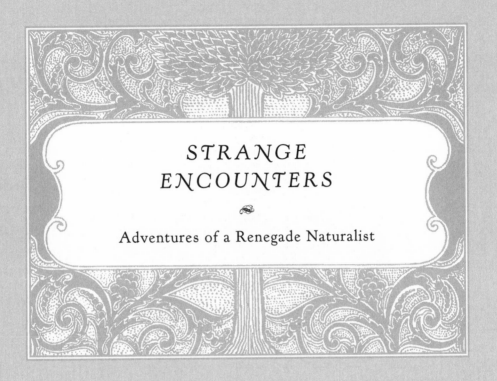

STRANGE ENCOUNTERS

Adventures of a Renegade Naturalist

DANIEL B. BOTKIN

JEREMY P. TARCHER/PENGUIN
a member of Penguin Group (USA) Inc.
New York

Most Tarcher/Penguin books are available at special quantity discounts for bulk purchase for sales promotions, premiums, fund-raising, and educational needs. Special books or book excerpts also can be created to fit specific needs. For details, write Penguin Group (USA) Inc. Special Markets, 375 Hudson Street, New York, NY 10014.

Jeremy P. Tarcher/Penguin
a member of
Penguin Group (USA) Inc.
375 Hudson Street
New York, NY 10014
www.penguin.com

The author gratefully acknowledges permission to quote lyrics from "Roll On, Columbia," words by Woody Guthrie, music based on "Goodnight, Irene" by Huddie Ledbetter and John A. Lomax. TRO © Copyright 1936 (Renewed), 1957 (Renewed), and 1963 (Renewed) Ludlow Music, Inc., New York. Used by permission.

Library of Congress Cataloging-in-Publication Data

Botkin, Daniel B.
Strange encounters : adventures of a renegade naturalist / Daniel B. Botkin.
p. cm.
ISBN 1-58542-263-0 (alk. paper)
1. Nature conservation. 2. Human ecology.
3. Ecology—Research—Anecdotes. I. Title.
QH75.B677   2003
333.7'2—dc21   2003050118

Printed in the United States of America
1   3   5   7   9   10   8   6   4   2

This book is printed on acid-free paper. ♾

Book design by Stephanie Huntwork

IN MEMORY OF JANE MITCHEL O'BRIEN,
who loved life and lived every minute as if it were the last,
enjoying each moment for all that it offered. Her enthusiasm
and love helped me pursue and enjoy the writing of this book.

ACKNOWLEDGMENTS

My editor, John Schline, Vice President and Corporate Director of Business Affairs at Penguin Group (USA) Inc., made many useful suggestions about this book, and was always a source of enthusiasm and support for the work I did. Our collaboration began when I brought him a group of stories; he worked solidly with me to structure and organize them to tell the precise narrative I had intended. Among his wonderful recommendations was the subject of the last chapter, which came up during a Sunday walk in New York City when I told him the story.

Special thanks are due to John Bockstoce, Lee Talbot, and David White for letting me share their stories and use their names, and to Richard Pfilf for his cheering story about Maggie's Bend. I also thank Lee for reading much of the manuscript and providing his usual helpful insights and comments. My now deceased father-in-law, Heman Chase, who taught me about old mills, New England forests and their charming people, is a voice behind all these stories through his strong influence on my life.

I am indebted to all the people whose personal accounts I have told, whose identities I have changed to protect them, for living their lives and providing the rich material that made this book possible.

Finally, I'd like to thank Diana for finding me after all these years and helping me when I needed it most.

Daniel B. Botkin
San Francisco, California
March 31, 2003

CONTENTS

STRANGE
ENCOUNTERS

"I give up. My father will never stop traveling," Nancy, my daughter, said over the phone to Jane O'Brien, my spouse. Nancy was trying to locate me. Like most of our relatives and friends, her first question when she called was, "Where are you?" Traveling has been part of my life. I have worked for thirty years trying to understand how nature works and trying to help solve environmental problems. Much of that has involved travel of some sort.

I have tried to live a reasonable life, to bring my kids up well, and to get along as best I could in the world. And I have tried to have fun along the way. In the process of doing my work and living my life, I have had some curious experiences. Often, the simplest facts that I thought would be easiest to find eluded me. Sometimes people would get into debates and arguments that didn't seem to make sense and to hate each other unnecessarily. Sometimes people did things that were funny. I learned a lot from the human side of nature and wilderness.

These experiences raise several questions. What can people expect of science when it is applied to the world immediately around them? What is the best that such science can do? What can we realistically expect of science when it is done by human beings, with all our failings as well as our strengths? What can scientists reasonably expect from the people, governments, and other institutions charged with overseeing, managing, and improving our surroundings?

When I give public talks, I often use one or two of these stories. People seem to enjoy them. Each has a point; sometimes the point has to do with our inability to think clearly and the peculiar problems this gets us into. This book contains some of my stories. Where I have worried that I may have inadvertently offended someone, I have fictionalized their name and affiliation.

A change is taking place in our myths about nature—in our nature knowledge—with immense implications for ourselves and life on the Earth. I am not talking about the Armageddon feared from human destruction of the environment; that is another story, related and unrelated. This is a change taking place subtly but definitely, as soft as a kitten but with the momentum of an elephant. It is perhaps the greatest change in human perception of nature and our role in nature in the last four thousand years—since the beginning of the great myths of Western civilization about the balance of nature. The change confuses us and leads to poignant and humorous incidents that are the symptoms of the change—the inflamed pustules on our body of nature knowledge. This shift has taken place during my lifetime, from the 1960s to the present. Working in ecology, I saw how it affected people and how their actions were perceived. It is best illustrated in the stories that happened to me or to one of my friends, beginning with the story of Maggie's Bend.

One

MAGGIE'S BEND

aggie's Bend was a whorehouse on the Clearwater River near Kooskia, Idaho. One day the feds arrived in town and announced that they were naming the Clearwater River—from its source down to the bridge just above the town—to the National Wild and Scenic Rivers system. They were acquiring scenic easements and purchasing private land along the river, including the part near Kooskia where Maggie's Bend stood. At the time, prostitution was legal in Idaho, and the U.S. Forest Service was given the task of establishing a fair market value for the property.

Maggie's Bend was owned by a fellow known as Whorehouse Jack. Few in the vicinity knew his complete name—or real name for that matter. His house contained nineteen fully carpeted bedrooms and was worth quite a lot, especially compared to typical houses in that part of Idaho. The facility presented the feds—here the U.S. Forest Service—with a special dilemma: How could they arrive at a fair market value for the business to assure that Whorehouse Jack would be fairly compensated? The Forest Service had to establish the value of the capital investment *and* the value of the ongoing business, based on present income and expenses as well as past and potential future profits.

The straightforward method of visiting and using the facilities and determining their value firsthand—in this case watching the goings-on and the exchange of funds—wasn't quite the right approach for an employee of the U.S. Forest Service. What to do?

Some locals suggested that the house ought to be kept within the Wild and Scenic Rivers system anyway, since it met with their ideas of wild and scenic and would probably be welcomed by many of the men, if not the women, who struggled down the Clearwater through the wilderness until they reached Kooskia, which was near the southern end of the planned wilderness area.

The supervisor of the Clearwater National Forest called in his real estate specialists to figure out a solution. They proposed visiting the laundry facilities in Kooskia used by Maggie's Bend and estimating business activity by counting the number of towels washed each day: a clean solution to a touchy situation. Of course, the real estate specialists had to know the number of towels used per transaction. This number would likely vary with each transaction, so the ratio would have an average value with an estimate of variation—what statisticians call an "estimate of error." Some moralists would call the whole venture an estimate in error. My sources do not tell me how the supervisor or his staff arrived at this ratio, or if this method was actually used, but we can assume that he did so by some legitimate means—statistically speaking, of course.

The forest supervisor's solution to Maggie's Bend has a certain charm to it, especially to me, because I have spent my professional life trying to help improve the environment and to understand how the interdependency between people and nature works. Often I find that people become so upset by the mistakes and failures in rationalizing our approaches to conserve nature that they become puritanical. They speak with a sense of revealed truth about nature, a religiouslike faith that they know what the Truth is about an issue. Instead, they might learn how to count the towels.

I was reminded of Maggie's Bend when I went to a speech by John Major, the head of Softsound, a Fortune 500 corporation that makes ceiling tiles. They are made in small squares so that damaged ones can be easily replaced.

Major was a good speaker in the way of an old-time revivalist preacher. He began his sermon by having everybody stand and hug his neighbors in the audience. Then he described how he had read a book that he said was called *The Death of the Earth* and had, as he phrased it, experienced an epiphany. He said he realized he had been an environmental sinner. In a flash, he got environmental religion and decided to change. Major had become an environmentalist.

"We're all environmental sinners," he said. "Join me, my fellows, join me. Won't you all admit that you are environmental sinners? I ask all of you to stand up and admit you are environmental sinners."

We all stood up and admitted that we, too, were environmental sinners. I wasn't going to be the only person sitting down in that crowd. Why, I was as good an environmental sinner as the next person.

Then he went on to tell us about the new environmental approaches he had had his corporation take. Softsound had decided to no longer sell ceiling tiles or other plastics but lease them and use returned tiles as the material to make new. It turned out to be much, much cheaper to make new tiles from old material than to purchase new material. The fact was that the new practices Major instituted were just good business.

The rest of us environmental sinners were supposed to atone for our sins, he explained, by increasing the efficiency with which we used resources. We had to become environmental puritans, as I understood his message, ever doing with less, ever increasing our efficiency, ever admitting our environmental sins, and never enjoying ourselves. We weren't assured of a huge increase in revenue, as his corporation experienced, but we would be better people.

I found this sense of revealed truth about environment scary, because it suggested that there are some people with the real truth and others without it. In my imagination, I could see people in white sheets burning wooden crosses labeled "Environmental Sinner" on the lawns of houses of people who wouldn't stand up and admit their environmental sins. It reduced truly complex problems about incredibly complex systems to simple answers.

The talk about being sinners made me realize what was so refreshing about Maggie's Bend and Whorehouse Jack. The events in Kooskia were about legally designated sinners confronted with the development of a legally designated wilderness. Parties on both sides sought a means to reach a fair agreement, and they did. To my knowledge, nobody had to stand up and admit who was and wasn't a sinner. In its own way, Maggie's Bend seems to better represent the truly wild, the truly natural, as well as the truly human, much better than the business executive who had increased his bottom line by many millions of dollars, but phrased this action in terms of righteousness and sin.

Nature is many things, but it is not hypocritical, and it isn't cast in black and white. There are not good species and evil ones. The wild is wild, make no mistake about it. It is messy, complicated, hard to hike through, and hard to understand. It's dangerous and might kill you if you're not prepared and alert. Sometimes it's even fun. It's about as difficult to comprehend as somebody counting the towels in the laundry at a whorehouse in Kooskia, Idaho.[1]

Two

THE RADIOACTIVE FOREST

Things are in the saddle and ride mankind.
—RALPH WALDO EMERSON

 n the 1960s and '70s, Brookhaven National Laboratory on Long Island, New York, conducted a unique, curious, and now little-known experiment: The laboratory radiated an entire forest. Back in those Cold War days, the danger of a nuclear war and of other releases of radioactive materials seemed real. People were building bomb shelters against the "Big One." High schools showed students films about what to do when the rockets went off—and what the world might be like afterward. The Navy sank a fleet of World War II ships by blasting them with an atomic bomb off the Pacific coral atoll called Bikini. Scientists were becoming concerned about the effects of an atomic war on the environment. Would we poison the world so that, if we did not die immediately in the initial blast, we would die slowly because crops would not grow and forests were dead? Would all the world's mammals, birds, insects, and plants die off, leaving us on a colorless, soundless, empty Earth that would slowly do away with us? In effect, World War II explosions of atomic bombs at Hiroshima and Nagasaki, the Cold War, and Sputnik had created the radioactive forest.

I was one of the researchers in that radioactive forest. It was my introduction to big-time environmental science, long before and far away from Kooskia and environmental epiphanies. We could work in the radiated forest four hours a day because the radiation was the relatively "clean" kind—from the heavy metal cesium's radioactive isotope 137. This produced only gamma rays, like X rays only with much shorter wavelengths and much deadlier. The laboratory moved the largest source of cesium 137 that could be safely handled by earthmoving machinery into the forest and mounted it on a vertical, movable pole with gadgets that allowed the radioactive material to be lowered into the ground and protected under lead shielding four hours a day. The pole held the cesium high so that it could spray gamma rays in all directions onto all the life in the forest.[2]

Two layers of chain-link fence surrounded the radioactive forest. One day, the forest's main research technician, Dave Whitcomb, and I went out to the forest. The sun shone intensely, creating a humid haze in the dizzying summer heat. We stopped at the outer entrance to the forest, a gate in the first of the two chain-link fences. A truck semi-trailer, which served as a field laboratory, marked the spot. We parked by a tiny pond in which small tadpoles swam busily. Insects buzzed. An occasional mosquito bit. Ovenbirds called. Sweet fern, a common flowering plant covering patches of the ground in the forest, emitted a pleasant spicy scent. The woods were deep green.

Seen from a low-flying aircraft, the eastern part of Long Island was heavily forested and looked a dark greenish brown—from the bark and needles of pitch pines—with a scattering of dark greens from the thick leaves of scarlet and white oaks and many shrubs, including blueberry and huckleberry. Brookhaven National Laboratory stood near the very eastern end of Long Island, a landscape shaped like a lobster's claw, whose northern finger was called Orient Point and the southern, Montauk Point. At the inside of the claw, the westernmost reach of a long bay, lay the town of Riverhead, and just inland from that was an airfield and production facility of Grumman Aircraft. Just inland from Grumman, in turn, was the laboratory.

Continental glaciers formed Long Island about ten thousand years ago. Like Cape Cod to the north, Long Island's northern half was part of the terminal moraine of the glaciers—material bulldozed by a mile-high blade of ice and left at the end of the retreating ice cap. Farther south, the island was outwash, material moved by rapidly flowing meltwaters as the glacier receded.

Moraine, made up of all sizes of soil particles pushed by the solid ice, was fertile. Outwash, primarily sand sorted from other sizes of particles by the speed of the flowing water, was infertile. On the infertile outwash, where the laboratory stood, grew a kind of miniature forest, dominated by pitch pine, white oak, and scarlet oak, none of them reaching much more than forty feet into the air. Below grew a dense cover of shrubs, most in the blueberry family, along with sweet fern Dave and I had smelled and other pleasantly aromatic herbs.

At the time, few people found this flat land with its miniature forest attractive. It was not a major location for hikers or other nature enthusiasts, but it had its special charms, with its aromatic scents and songbirds calling.

We parked next to the trailer, opened the outer gate with a set of massive keys, and walked past a smaller car-pulled trailer that was full of scientific equipment from which came loud pounding noises. Beyond that trailer was the inner fence, and we opened its lock. From here on, when the radioactive source was up, radiation was higher than the background that exists everywhere on the Earth—a background radiation that we all receive, the result of cosmic rays raining down on our planet.

Just inside the inner gate, the forest was still dense and deep green, a thicket of shrubs and small trees, just as it was outside the gates.

We walked along a well-used path toward the source of radiation—toward ground zero. Above us, strung from metal poles, ran miles of clear plastic tubing and black electrical wires, all part of experiments within the forest. Every ten meters a sign on an aluminum post announced the number of meters from the radioactive source. As we walked down the path toward the source, I saw a decline in the vegetation. First, pitch pines had dead branches and twigs, then some were dead, then all the pines were dead. Then the same pattern with white oak. A few minutes more walking and the scarlet oaks were dead. No trees survived past this point. Only small shrubs, like blueberry, huckleberry, and sweet fern, intermixed with small grasses and sedges, remained green.

A few minutes later, walking now in the bright hazy sun closer and closer to the source, with no tree leaves to shade us, I saw that most of the vegetation was dead. The only live plants were those in the shadow behind standing dead trees. These small plants grew in a triangle that faced outward from the source, so that there was a circle of these triangles around the source, each pointing directly away from the source. The gamma rays, like any electromagnetic radiation, lit up the forest with intense radiation, but were absorbed by the dense wood of the larger trees. The exact triangular shape of the patches of grasses and shrubs was an eerie demonstration of a basic principle of the physics of light radiation.

As we approached within fifty feet of the source, we saw that a few scrubby triangles of sedges had persisted behind a few large, dead trees. Then there was a totally dead area perhaps ten meters—thirty feet—in radius. And there, at the center, was the rectangular top of the shielding: ground zero.

Cesium 137 emits a "clean" form of radiation—gamma rays that are dangerous but leave no residue. When the source was underground, the forest was nonradioactive. There are two other kinds of radiation that are "dirtier," called alpha and beta rays. Alpha rays are nuclei of helium atoms flung out from radioactive isotopes. Beta rays are electrons, also flung out from radioactive isotopes. These are "dirtier" because when the original source is removed, the radioactive particles remain and can continue to do damage. They can be absorbed by a person's body, or the roots, stems, or leaves of a plant. They can hide in the soil to be picked up and ingested at some later time by a mouse, a deer, an ant, or some other kind of life. One could not walk through that kind of irradiated forest, full of alpha and beta particles, in street clothes. You would have to wear a kind of space suit with an internal air supply and lead-shielded clothing head to toe.

Another experiment, at Oak Ridge National Laboratory in Tennessee, put alpha- and beta-emitting radioisotopes into ecological food chains by injecting them into the bark of trees so that they were transported up to the leaves; when the leaves, twigs, and bark from the trees fell, they were deposited in the soil and eventually ingested by animals. The scientists traced the pathway of radioactive isotopes within the forest food chain, similar to the way doctors use radioactive iodine and other radioactive chemicals to trace pathways through your body.

A third experiment took place in the rain forest of Puerto Rico. Its purpose was to study the effect of a short-term exposure to gamma rays, called an "acute exposure." A large source of cesium 137 was flown into the forest by helicopter, left for a short time, and then flown out, so that the recovery of the forest from this brief exposure could be studied. This contrasted with the forest at Brookhaven, where scientists studied the effect of long-term, persistent radiation—called a "chronic effect"—on a forest. The Cold War had led to hot forests.

The view from ground zero, standing on the lead shielding on top of the cesium on a July day, was eerie and unworldly. Dead trees, mostly standing, some fallen but not decayed, surrounded me. Their bark appeared somewhat blackened, as if a fire had just passed through. The gamma rays sterilized the forest, killing bacteria and fungi, earthworms, ants, and the other forms of life that decay wood in a forest. It was a dead forest frozen in time. If you had not known about the cesium source, you would have believed that the forest had burned a few days, perhaps a week before—not so recently that branches still smoldered and smoked or that every step and every breath of wind raised fine ash, but recently enough so that most of the dead trees were still standing. I wondered how safe it really was to be standing there in this eerie scene of death and undecay.

Technical experts on radiation, known at the time as "health physicists," told us that it was perfectly safe to walk around the forest when the cesium 137 was buried and shielded. But the laboratory's health physics department also required that each of us wear a little white badge, about the size of a large postage stamp, any time we entered the forest. Each badge held a strip of photographic film that responded to gamma radiation and indicated if the wearer had suffered any exposure to radiation; it also measured the amount of exposure, including lethal doses.

Putting the two statements together—that the radioactive forest was perfectly safe to work in as long as the source was in the ground, but that we had to wear radiation-detection badges—I decided to test for myself how much radiation actually came through the lead shielding; on this day I carried a Geiger counter. I guess you could say that I had a fundamental distrust of large bureaucracies and, more to the point, was the kind of person who had to prove everything for himself.

I had turned on the Geiger counter before we entered the outer gate, read its needle-faced instrument, and listened to the staticky sound of cosmic radiation striking the detectors in the instrument. The needle bounced up and down, but remained around a low value, which I wrote down in my field notebook; this was the natural background radiation level.

At ground zero, I walked onto the lead shielding and put the Geiger counter down next to me, right above the cesium 137, protected from it only by lead blocks. I knelt down and read the dial. Exactly as we had been told—the needle bounced around the same numbers it had outside the outer fence. The nuclear engineers who had designed the shielding knew what they were doing. I had proved to myself that it was safe to be there. I hoped that the then new big science and big money would allow us to do as well in understanding forests as those engineers and health physicists understood radiation. Why not? We were all scientists. We were all pursuing truth.

I turned off the Geiger counter, put it down, and looked around. Standing at ground zero, seeing this frozen devastation, I wondered how visitors from other times and other civilizations would perceive this forest. Suppose in some future time, a world after the Big One, the great atomic war, some hunter-gatherers would come upon this place, having climbed over the remains of its fences and reached the center where the source, with its lift mechanism no longer functioning but still standing, still irradiated the forest—albeit at a lower intensity, because the source slowly gave up its radioactivity over time. Would they believe they had come to some sacred grove, where the cesium on the pole represented the icon of some woodland god? Would they believe that the devastation that would not disappear, would not decay, was a sign of the hidden, mysterious power of that deity? Would they pray and carry out some kind of ritual to appease this woodland deity so that its powerful poison would not reach beyond the fences? Would those who visited become sick and perhaps die, adding further to the belief in the power of the god of this sacred grove? Would they try to destroy the idol, knock down the metal on top of the pole? Or would they worship it, sending it slaves or captives to be sacrificed?

Or suppose the Big One never happened and the radioactive forest was forgotten, a hidden corner of a brief period in twentieth-century science, only to be discovered several centuries in the future by some rambling boys and their dog, and reported back to the scientists, scholars, and politicians of that future time. Would they scratch their heads trying to figure out why any rational beings had intentionally created a forest that was not only dead but sterilized? Would they think that ours was a most peculiar civilization that seemed to have gone off on a kind of crazy tangent?

But the heat of the day soon took my mind off these musings and I began to think about the work that I had to accomplish during the few hours that remained with the source underground.

Having majored in physics as an undergraduate, I was comfortable with the principles of electromagnetic radiation and with the calculations required in this study. But I hated being stuck inside a laboratory and welcomed the opportunity to participate in a study outside, within nature. I happened to join this study just as the first big money came into it. That had not been my expectation nor my desire. I simply wanted to participate in the study of nature, to be outside as I did that study, and to try to help improve this field as a science, however I might. Standing at ground zero, I did not yet know how that might be. Nor did I understand the status of this new field, ecology. Trained in the formalisms and rigors of physics, I had not yet considered that ecology might be any less rigorous than those fields. After all, ecology was a science, wasn't it? Weren't all sciences alike? We were living in the age of science and technology, and I was fortunate enough to be one of the participants in one of science's newest fields.

All the other work I had done in my as yet short career in ecology had been low budget, on a shoestring: a walk in the woods with some surveying equipment, measuring the distances between plants; empty tin cans buried upright in the soil to make traps to catch insects, a way of sampling that kind of fauna in a forest; simple pencil-and-paper mathematical equations that were supposed to provide a theory for ecology.

Dave Whitcomb and I began our work. We were repairing some broken wires and plastic tubing, and I needed to cut some tubing.

"Can I borrow your pocketknife?" I asked Dave.

"Why don't you get your own?" he asked. "They have them in the biology building supply room." He handed me his.

"I don't believe it," I answered. "No government agency is going to give out pocketknives. People will just put them in their pocket and go off with them. The government would run out of them."

"Ask them," was all that Dave would say.

That afternoon, when the source had risen back above the ground and we were well away from the radioactive forest, I returned to the biology building and immediately went upstairs to the supply room. It was crowded with the kind of metal shelves you see in a warehouse, painted gray, put together with big nuts and bolts like an old-fashioned erector set. Each shelf was filled with equipment. One held plastic laboratory ware. Another, mechanical parts. There was everything a biologist might need in a hurry but rarely had easy access to. Crowded in one corner, at a desk piled with paper, sat Mike, the supply officer—lean, a thin face, back slightly bent. He ruled over his dominion with a firm hand, like a master sergeant, and he was a little frightening to a beginning graduate student. His manner was gruff, and he stood between me and all the necessities of my research. Still not believing Dave, I looked around the shelves for a while to try to find a pocketknife without actually having to talk to Mike. But I saw none, so I summoned up my courage, expecting to be laughed at, and walked over to Mike's desk.

"Mike," I said, "I would like a pocketknife."

"Which kind?" He replied, "Swiss Army or electrician's?" I took the Swiss Army.

By visiting the supply room several times, I soon had *two* pocket knives. This was my introduction to the world of post–Sputnik, highly funded, high-technology scientific research. Ah, the bounty of the post–

Sputnik scientific race between the United States and Russia: the wonders of equipment suddenly available to this humble field of study—two
beautiful and different pocket knives, free, courtesy of the American tax-
payer and his fear of nuclear war and losing the space race to the Russians.

If I wanted paper, I was driven to the paper warehouse. If I wanted a
roll of electrical wire, I was driven to the electricians' warehouse. The same
for wood, metals—any kind of equipment. These were big warehouses.

Yes, only the best equipment was allowed on the Brookhaven site.
After all, it was the location of some of the world's largest research
nuclear reactors and some of the most advanced research in physics at
the time. Our little ecology project, hidden off in a corner of the land,
was small potatoes, with relatively small funding compared to the giant
reactors.

The attitude of the staff at Brookhaven, that only the best equipment
could be used, was brought home to our small ecology group one day
that summer when a graduate student wanted to sample insects in the radioactive forest. As mentioned before, one standard sampling approach
was to take an empty tin can and bury it in the soil so that its open top
was level with the soil surface, trapping any wandering insects. This is
all the graduate student wanted, but it was unheard of to bring an old tin
can as a piece of research equipment to a national laboratory that was in
competition with the Russians. He was informed that he had to go to the
machine shop. There, he was asked for the specifications—height and
diameter of the cylindrical trap. A professional blueprint was made. A
week later, the device was ready: a stainless-steel beauty of a cylinder
charged to the ecology account for seventy-five dollars. A seventy-five-
dollar equivalent of an old tin can. It was beyond an ecologist's wildest
imaginings. Totally unnecessary, but a hint of the power of the money
that was readily available at that time.

Brookhaven National Laboratory represented the latest in high technology, but it also represented a large bureaucracy funded by the federal

government and therefore by federal income taxes. It was an extravagant approach to technology with abundant funding.

A few days later, Dave and I were sitting outside the research trailer under the shade of a scarlet oak, just across the outer gate from the radioactive forest, eating our lunch. Dave held up a small white rectangular plastic object, about an inch wide.

"See what I found?" he said. He held up one of the radiation badges that we were required to clip to our clothing when we entered the radioactive forest.

"It's just a badge. What about it?" I asked.

"It's mine. Lost it about two weeks ago," said Dave. "Found it this morning about two yards from the source. I'm going to hand it in." We were required to turn in our badges weekly and they were checked for radiation exposure.

I'm sure I turned pale. One day's dose at the source would kill a man. Dave's would register fourteen times the amount needed to kill him. I imagined Dave being hauled into the medical office in an ambulance, subjected to all sorts of embarrassing medical tests, and then being fired or put on some kind of probation for playing jokes with this serious business.

"That'll give them something to think about." Dave laughed. He was working at the lab in a break between college and medical school and thought that the entire work he did at the lab was unimportant—no more than amusing. To me it was the beginning of a serious career, but to Dave it was a way to while away the summer. He was good with machinery and scientific instruments and found the work easy to do.

While I shuddered to think what medical tests Dave might be forced into when the physicians got a look at his badge, Dave laughed again about the expressions he imagined would appear on the faces of the medical personnel when they read his badge.

It was a pleasant setting where Dave and I sat and ate lunch. A small pond was next to us, near the large semi-trailer used as a field laboratory.

Ovenbirds called again from the woods around us. The scent of sweet fern was in the air. It was midsummer and the sky was full of small cumulus clouds.

"Don't do it, Dave," I said. "It will get you into a lot of trouble."

"Trouble. I don't care. I'm leaving in a month, going to medical school. What do I care?"

Here we were, among the first to use the latest in technology for this ancient study of nature. In spite of Dave's lighthearted attitude about the project, I felt a thrill of excitement to be part of a cutting edge of a new science. There were never a great many people working in the forest at any one time, perhaps five or six at the most. Some repaired air pumps located around the forest to push air through the plastic tubes. Some collected insects. Others studied the soil to see if it was losing its organic matter. The noise of the pumps and the amount of hardware seemed to assure me that I was involved in big science and therefore *real* science.

After lunch we went back into the forest, walking down the main path. I looked at the wires strung overhead on metal poles. Fifteen miles of wires and tubing spread through the forest. It was probably the most wired forest ever. Wires carried electrical signals from sensors in the trees and soil, measuring temperature and moisture. Clear plastic tubes attached to the same poles carried air to and from clear plastic cylinders that surrounded leaves or stems on living trees. The air was measured for its carbon dioxide content, going in and coming out of these cylinders, providing a direct measure of the respiration of the stems and the photosynthesis of the leaves. In this experiment, with all the money available, the goal was not just to wait and watch things die, but to see if we could detect rapid changes in photosynthesis and respiration—in a sense to feel the pulse of the forest and see if it was slowing down. No one had tried to make these real-time measurements on such a scale before.

On our way back, we stopped at the small camping trailer that was the instrument headquarters, the brains of the research. It housed measuring and recording equipment. Dave had to do his daily calibrations of the instruments. A loud and rapid *chat-chat-chat* told us that all was well with the primary device, a paper-tape digital recording system—the first digital recording system ever used in ecological research. Yessir, this was top-notch science. A large instrument rack held a series of machines. One was a digital voltmeter—just invented—that put out a pulse that could be read in computer code. This instrument read voltages from a series of inputs and recorded the information as holes punched into paper tape. It was this paper-tape punch that made the loud noise. The paper tape was then taken to the main computer at the laboratory and the information on it read and converted to temperature, moisture, and carbon dioxide concentrations.

The trailer was packed full of equipment that had been put together by the machinists, electricians, and engineers of the laboratory. No one had made such a recording device before, so they had to be imaginative and use equipment that was available. There had to be a way to switch input signals from the many measuring devices into the one digital voltmeter. Some engineer had figured out that he could use a bank of old stepping switches obtained from an outdated telephone switchboard. This was an elaborate mechanical contraption. At regular intervals, an electrical clock sent a signal to a solenoid switch. This switch closed with a bang and moved the stepping switch one step forward, breaking the connection to one instrument and making the connection to the next.

On another instrument rack was an infrared gas analyzer, at the time a complex and delicate machine. It had two tubes, one with air with a known concentration of carbon dioxide, and the other with the air pumped in from one of the plastic chambers in the forest. Carbon dioxide absorbs specific wavelengths of infrared light (that's why it is a greenhouse gas when it is in the atmosphere), and a source of infrared light shone down each tube. At the far end, an infrared light detector measured the amount of light getting through and converted this to an electrical signal, which was then read by the digital voltmeter and represented a concentration of carbon dioxide in the air.

Unlike today's miniaturized integrated circuits, each device of this recording system was large and physically separate, connected by thick wires and tubes. It was fascinating to stand and watch the mechanical switches work automatically, the digital voltmeter's dial change rapidly with each new measure, and the paper-tape punch bang holes in a strip of paper tape about an inch wide. The tape spilled out onto the floor of the trailer. Later, Dave would neatly gather up the paper tape and bring it to the computer building. Punched holes filled the floor, so that the whole thing looked like an unkempt bachelor's apartment. Large fans and pumps roared inside and outside the trailer, some pushing air into and out of the forest, others cooling the air inside the trailer.

By twenty-first-century standards, this was an incredibly crude set of devices. Its total memory was far, far less than what is contained in a hand-held Palm computer, a mobile telephone, or a point-and-shoot digital camera, and probably less than in the little computers inside a modern automobile engine. But it was the first digital recording system used in a forest, the first used in the study of nature. Here, nature and computer technology were making their acquaintance. It was hot and noisy, but the trailer pulsed with what seemed excitement.

Standing at the trailer with its paper-tape hammer banging and its pumps roaring, I wondered again how other cultures from other times would view this forest. What would that hunter-gatherer band who came here after the Big One think of the wires, the clear plastic tubes, and the remains of the instruments in the trailer? What would a technologically advanced culture think coming upon this trailer?

Ecology is, in a sense, a formalization of natural history—the activity of observing, studying, and trying to understand all the fascinating forms of life around us. It was the twentieth century's attempt to try to answer some of the most ancient questions people have ever asked, have ever written about: What is the character of nature undisturbed by human beings? Why are there so many species and why are they so well adapted to their lives, their needs, and their environment? Does the entire system of nature function as a unit? And if so, was it always in a perfect and harmonious balance? What was our place in this nature, if any? Did we have a role, a niche, like all other creatures, in the great chain of being? Or, having been cast out of Eden, were we also cast out of that chain of being—outside of nature, not part of it, only a spoiler and destroyer of it? Why did nature change and why was it sometimes not perfect? This was a question that Lucretius had asked so poetically more than two thousand years ago in De Rerum Natura (On the Nature of Things). Of what benefit was this nature to us? Where were its medicines? Its sacred groves? Did it rule over our lives, and if so, how? Could we divine its intentions? How did we affect this nature?

Ever since people had written, they had asked these questions. Now, in the age of science, it seemed possible, to a small number of enthusiasts, that these ancient questions could be answered by applying the methods of big science, the kind of science that produced the atomic bomb.

By 1966, when I stood at ground zero in the radioactive forest, environmentalism had just begun to grab the public's attention. Rachel Carson's Silent Spring had captured the popular imagination. Three years in the future, in 1969, an oil spill on the California coast at elegant Santa Barbara would lead to the establishment of an organization called GOO for Get Oil Out, and to the establishment of the first major environmental studies program in the nation, at the University of California, Santa Barbara. Paul and Anne Ehrlich's The Population Bomb was two years in the future. The Limits to Growth (D. H. Meadows) study was under way but not yet published, providing a doom-and-gloom forecast of the future of the world. The major federal environmental laws also were in the future—the Marine Mammal Protection of 1972; the Endangered Species Act, to be passed in 1973; the Clean Water Act; the Clean Air Act.

The field of ecology had not yet come to center stage. Even among ecologists, the Brookhaven experiment was not well known. To those who knew about it, the experiment seemed strange and at the fringes of a science that had mainly focused on the study of the character of nature undisturbed by human influences. Here was a study of a forest inten' tionally subjected to a novel human influence. Some ecologists won' dered what this could possibly tell us about real nature.

Looking back at the equipment in the radioactive forest from the per' spective of today, less than forty years later, it seems incredibly crude. Few today would have any idea what that equipment did. Even elec' tronic, computer, and mechanical engineers would scratch their heads and have to run to old books and journals to figure out how these anti' quated and forgotten machines worked. They would no doubt find it amusing that telephone stepping switches, mechanical devices that would seem out of the Dark Ages, were at the heart of what made multiple measurements possible. Today, equivalent kinds of field measurements are made with much smaller and greatly more reliable machines. Hand' held computers take in data, locate the site of measurement using global positioning satellites, and radio data directly to a personal computer that installs the data in a geographic information system, where it is put in the correct geographic location. Layers of data of different kinds are auto' matically recorded in this fashion. Then a researcher opens a program and inputs these data into a graph'drawing program and out spit beauti' ful maps, graphs, and statistical analyses.

While Brookhaven seemed on the surface to be a place where people had technology under control, there were some suggestions that tech' nology might be gaining the upper hand. Sandra Sun, another graduate student, did a summer project examining how the laboratory dealt with its wastes. She learned that the laboratory was extraordinarily careful with the radioactive wastes, handling them exactly according to legal and regulatory requirements.

One of the standard regulations for the disposal of radioactive mate-rial down a sink drain was stated only in terms of concentration in the water, not in terms of the amount released. During the winter, I had taken a course in "radioecology." At the time, radioecology seemed to be part of the future of ecology, but in fact it became a sidetrack that almost no ecologist knows about today. In that class we were shown the proper way to get rid of water-soluble radioactive materials. We looked up the legally allowable concentrations of each radioactive isotope and then di-luted whatever we had with enough water to bring the concentration down to that legal limit. Then we poured the whole thing down the sink. We could put a large quantity of radioactive material down the sink, as long as it was at a legally low enough concentration. This was, as the modern saying goes, an example of "dilution is the solution to pollution." The belief was that if these toxic materials were diluted sufficiently, they would be safe out in the environment.

Meanwhile, Sandra discovered that the laboratory mixed together all nonradioactive wastes into one big brew; that brew was put through the laboratory's processing plant and into the soil of the laboratory. The soil under the laboratory was primarily sand and therefore highly porous; wa-ter and whatever water dissolves, including toxic pollutants, moved rap-idly through it. Because it is an island in the Atlantic Ocean, there are two distinct layers of groundwater in its sands. Seawater infiltrates the sandy soil. At the same time, rain and snow fall on the surface and seep into the ground. Since salt water is heavier than fresh, the two do not mix. A lens-shaped body of fresh water "floats" on the seawater, all of which is under the ground on Long Island. Both sewage and drinking water are connected to that lens of fresh water. Sewage goes into it and drinking water is taken out of it. Since the fresh water can and does travel horizontally underground, the two are connected. You have to hope that the sewage water is cleaned up before it goes into the ground, and that the transit rate is slow enough so that the wonderful community of living creatures in the soil—mostly bacteria—have time to work on what remains of the sewage and clean it up.

We students in the ecology group were quite concerned when we learned the results of the summer study of the laboratory's waste disposal. The soup of all kinds of chemicals might be a witch's brew creating new chemicals—certainly creating chemicals not in the original mixture. But the laboratory dismissed the report; Sandra's study was considered unacceptable and amateurish.

Dave left the laboratory at the end of the summer. Saying goodbye, he told me a story about another lab employee who had recently left. He said that employee came to the lab in his car and stopped by our offices to say goodbye to Dave. Dave went outside and saw that his friend's car was loaded down with research equipment he had casually taken from the various storerooms. So heavy was the equipment that it compressed the springs of his car and the fenders almost touched the tires.

The radioactive forest experiment ran for fifteen years—its design lifetime—after which the cesium was taken away and the forest abandoned. Curiously, no one did a follow-up study to learn if and how the forest recovered. Some interesting scientific papers resulted from the project about how radiation affected a natural forest. But this kind of study has gone out of fashion now that the Cold War is over, and few know of these papers. That these studies might be excellent examples of more general characteristics of pollution seems not to have remained on the ecology radar screen.

Dave Whitcomb, the first radiated-forest technician, never did take the research seriously. He seemed to think it was amateurish, not really science, a kind of joke and game. As we said our goodbyes, I had the troubling feeling that perhaps Dave was right, perhaps what seemed to be a state-of-the-art, cutting-edge, high-technology study of nature was really a joke, that we weren't doing science, just playing at natural history, and making the same old mistakes in the observation of nature that people had made for thousands of years—believing what we wanted rather than what data told us. Perhaps we were no different from those hunter-gatherers who, in my imagination, might arrive at the radioactive forest after the atomic holocaust and see it as an object of myth, mystery, and worship. Perhaps our high-technology devices and access to expensive equipment merely fooled us into believing we were doing science when we really were not. Perhaps the radioactive forest was at best a good joke, a joke on science, a joke on ecology, a place that, to whatever kind of civilization follows us, will seem strange.

In the years since I said goodbye to Dave, I have become involved in many other strange, sometimes funny projects that continued the attempt to understand nature. Have these been a search for truth or no more than what the fool told King Lear, a nothing from which could come nothing?

Three

*

REPAIRING THE MILL

uring the time that I was working in the radiated forest at Brookhaven National Laboratory, I spent vacations and weekends in rural Alstead, New Hampshire, helping my father-in-law, Heman Chase, a country surveyor, with his work and his hobbies. His main hobby was a working water-powered mill, the best of nineteenth-century technology. Traveling back and forth from Brookhaven National Laboratory to southeastern New Hampshire, I was moving from one century to another, and the experience was often jolting. But more than that, I was traveling between two entirely different views of life. Heman believed in independence and in a small, self-sufficient community that used its natural resources well and depended as little as possible on trade outside. He and I shared a love of technology and of nature. He was a strong opponent of big government—the kind of government represented by Brookhaven National Laboratory—and I doubt that he ever understood why I was working in the radioactive forest set up by a big federal agency. How could I reconcile these two lives, I wondered.

"If you believe you should give a dollar to the government, then you have to believe that it will spend it better and more efficiently than you would or somebody you know would spend it," he said to me one day when we were surveying an old farm grown back to woods. We spent many days doing such work, surveying land that was passing from open fields to forests and from farm and pasture to the homes of retired executives or to vacation homes of the wealthy. We enjoyed the forests, wetlands, rocky slopes, and hills of New Hampshire and Vermont through this work. As we worked, we discussed politics, philosophy, economics, and just about everything about people, society, and nature. He believed in the independent life in which a person did good works and appreciated all he could about civilization and nature. Heman patiently taught me the names of the forest plants as we surveyed our way through the forests.

I came to the southeastern New Hampshire countryside at a time of transition. It was still possible to make a living as a dairy farmer with as few as fifty milking cows. A decade later, that lifestyle would end with the introduction of new sanitary regulations requiring that all milk be produced with electric milking machines, pumped directly into stainless-steel tanks, and from these offloaded into stainless-steel tank trucks, never touched by human hands, never in danger of contact with the dirt and manure of the milk shed or the barn. It was as if the high tech of Brookhaven National Laboratory came to independent, rural New Hampshire.

At that time, southern New Hampshire was little known. It was out of the way for the major summer tourists who took the cable car up Mount Washington in the Presidential Range of the White Mountain National Forest to the north; out of the way of huge Lake Winnipe-saukee in the north-central part of the state. Summer visitors to Alstead were mainly city relatives of year-round natives or couples who had taken early retirement and chose to escape the urban pace of New York or Boston. Some found ways to enhance their incomes doing odd jobs.

To outsiders, the old-timers of this remote region were known to be taciturn but to have a special, dry sense of humor, one that seemed from a different world. Among themselves, they socialized by speaking little but saying much. Walter Burroughs, Heman's brother-in-law, was typi-cal. One cold winter morning in Alstead, I woke early and decided to go out to get the mail. As I approached the row of mailboxes mounted about chest high on a six-foot length of a 4-by-4—the community mail center—I saw Walter Burroughs approaching from the brick house where he lived. Walter was a country gentleman. Lacking formal education, he had a graceful manner and an ability to put others at ease. He was in his eighties and well known to have been, and to be, a heavy drinker—what we would call an alcoholic. Walter was a quiet drunk, never bothering anybody as far as we could tell.

"Good morning, Walter," I said. "How are you?"

"All right," replied Walter, "for a man of my age and habits."

One day later that winter, Heman and I were down in the bottom level of his old water-powered mill, replacing the flume—the long pipe that carries water from the millpond to the waterwheel, providing the power to drive the wheel. We were immersing ourselves in nineteenth-century technology. The old mill stood beside the stream and near to the road at the top of a rise. It had an unmortared foundation of three-foot-high glaciated stones. The mill was fronted by weathered, unpainted ver-tical pine boards—one of the picturesque architectural styles of rural New England. A cold wind blew in through the cracks in this founda-tion. Water from the millpond leaked down past us, cascading onto a layer of ice and skidding down the millrun out to the opening in the foundation and into the main channel of the stream. Although called the basement of the mill, it was really a construction over a millstream cut partway into a slope.

Near the upstream wall where the flume came out of a hole in the foundation, carrying water from the millpond to the wheel when everything operated correctly, Heman Chase struggled to shove a length of corrugated, coated iron piping about three feet in diameter into place; we would then fashion metal clamps around the length of pipe. Heman and I pushed and shoved this section of the corrugated piping, as we had many others, into place. It was exhausting work, and it always seemed colder inside the unheated building in the New Hampshire winter than outside.

I thought about the laws of physics that told me why this was so. Each of us radiates about as much energy as a 100-watt lightbulb. When we are in a warm room, the walls radiate heat back. When we are in an unheated building, the cold walls absorb that heat faster than we can generate it, faster than it would be lost outside in the sun with no wind blowing.

Air is a pretty good insulator, and you're warmer outside in winter on a sunny day than inside the wet basement of an old stone-foundation mill. But that knowledge gave me little comfort. My hands felt frozen against the corrugated metal. My feet were almost numb from the cold, and my nose and cheeks felt iced. We had to stretch and turn our bodies in awkward positions to move the big pieces of metal. Sometimes we did this lying on cold rocks; sometimes ice and water chilled and wet us.

The water that flowed past us as we worked leaked from a ten-foot-high millpond dam just outside the foundation. The millpond was fed by water flowing from a body of water called Warren's Pond, a few hundred feet away. In typical New England understatement, this "pond" was what elsewhere in the United States would be called a lake, extending three-quarters of a mile to the north and spreading out into several coves. The pond's own dam, about four feet high and ten feet long, the millpond dam, and the mill building had been there for more than a century.

From Chase's mill, the stream descended a steep grade, twisting through a glaciated valley. Walking along the stream, you could find the foundations of six other mills that had once been powered by the same water, all within a quarter mile of the remaining mill. Here was one of the small power centers of nineteenth-century New England, part of the Industrial Revolution within this most independent village of this independent state. Heman believed in that kind of independent industry, but I doubt that he would have stood up in a revival meeting about the environment, admitted he was an environmental sinner, and tried to use this to increase his bottom line.

"How many times have you had to replace the flume?" I asked Heman.

"Let's see," Heman said. "My stepfather, Hartley, and I began to restore the mill in 1917 when I was a teenager and we put a new flume in then. Since then, I think I've replaced it twice more. Don't expect I will need to do it again." Heman was in his sixties, took digitalis each day for his heart condition, and ate a diet heavy in cream, ice cream, butter, and rich meats.

"Time for a break," he said after we finished installing one of the metal sections. We uncoiled ourselves from our awkward positions, backs and legs against the hillside of the mill basement, and half-crawled from the upstream end of the building toward the waterwheel. The flume, when intact, extended about forty feet horizontally and then made a vertical turn downward so that water flowed down a pipe and into a turbine waterwheel. This wheel, shaped like a fan and similar in design to the blades in a modern jet engine—or, on a much larger scale, a modern water-powered generator—was a relatively new invention in the second half of the nineteenth century. All the water falling through the flume down ten feet made contact with all the turbine blades all the time, so that much more energy was extracted from the water than from a wooden overshot or undershot waterwheel seen in nostalgic landscape paintings, whose paddles only touched the current part of the time.

Having crawled and walked away from the beginning of the flume, we went down a rickety wooden stairway to the level of the wheel and inspected its condition. The outer housing of the turbine blade was partially encased in ice, but the external metal seemed to be in good shape. The wheel was fastened to a vertical metal shaft, painted yellow, that transferred the rotating motion of the turbine to the floor above where Heman and I had been working. We walked up the rickety stairs and examined the steel shaft where it was attached to a large pulley, about three feet in diameter. When the flume was intact and carried water to the wheel, this pulley pulled on a leather belt, about eight inches wide, that spun a horizontally mounted pulley about twenty feet away. The second pulley was fixed to a long, horizontal metal axle that ran almost the entire length of the basement, perhaps thirty feet, and on which were four other pulleys. Each of these spun other leather belts that transferred power to yet another pulley at the base of another vertical shaft. Some belts were horizontal and moved the rotating power across the basement to other pulleys; other moving belts went up through the ceiling and turned woodworking tools.

When the millwheel ran, everything spun, thumped, and moved, and the entire building vibrated—all five floors of it, if you included the partially open basement; one floor for the waterwheel; one for the belts and pulleys to transfer power; one for the machines and wood- and metal-working shop; and living quarters above, if you could keep warm enough in the winter from its one fireplace and plank-board, uninsulated walls. You had a sense that you were in direct contact with energy and power. It was a plain-to-see giant demonstration of some fundamental laws of physics and rules of engineering, so different from the hidden electronics of machines in the radioactive forest, or for that matter the radiation itself—invisible, deadly, sterilizing. Here the vibrations threw dust of two generations of users into the air. The dust reflected and refracted sunlight that fell through the windows. Energy was made suddenly visible.

"Best way to teach a child about physics," Heman said. "You can see the energy transferred from waterwheel to pulley, from pulley to metal shaft, and from metal shaft to a machine." On weekends he ran a shop class for local children, introducing them to woodcraft and to the laws of physics through a thorough tour of the millwheel and its belts and pulleys, as well as of several cutaway working models of mills, one of which ran from a tiny spray of water from an offshoot of the flume. I agreed that this entirely mechanical transfer of water power was a clear example of nineteenth-century industry and energy production.

We drank some hot cocoa out of a Thermos and ate a few cookies and then resumed our work, getting a second section of the flume in place by the time the sunlight was fading. Then we stopped for the day and went up the road to Heman's home for a cozy dinner made by his wife, Edith.

We were back at the mill the next day, on the main floor in the wood- and metal-working shop. It contained some electrically powered tools, many hand tools, and four machines powered by the water: a drill press, a joiner, a power saw, and a planer. The primary use for the waterpower was to plane rough boards. The waterpowered machines were also of nineteenth-century manufacture and were well made, as was characteristic of that period in America. The planer took a rough board that would put splinters into your fingers if you handled it without gloves and spit out wood chips and a beautiful smooth-surfaced board. It was a mechanical marvel in itself—about four feet high, four feet wide, and six feet long. Its blades were a series of rectangular chrome-steel sections put together on a belt, like treads on a bulldozer. They spun so rapidly that they seemed a blur, making a very loud chattering roar when a board passed through them. It took some skill to use the planer. If you pushed up or down on the board as you tried to push it through, the blades would gouge the board and make it unusable. The noise was deafening. The planer grabbed the board from me and pulled it with great force; using it was another way I felt in direct contact with energy and power. Heman said that the water turbine generated fourteen horsepower, and you could feel every one of those horses yanking the board away in the planer.

A small woodstove burned brightly in the shop, warming the main floor comfortably even on this harsh winter day. As I had done many times, I looked around the shoproom of the mill. A half-dozen or so workbenches filled the spaces between the big power tools. Every nook and cranny was crammed with hand tools of the nineteenth and the twentieth centuries, along with partially finished projects—a child's model airplane, a three-legged stool, a reproduction of an iron-banded carriage wheel partially assembled—jars of screws, nails, fasteners, and strangely shaped, elegant brass objects whose functions were unknown to me. There were hand saws, chisels, clamps, and drills. An old banjo clock hung in one corner, a dusty antique little noticed, later assessed to be worth thousands of dollars. I suppose that many of the other objects stacked in odd corners here and there and also covered with dust were equally valuable and would have been a joy to a museum of technology. Many of these objects had been there for fifty years. They were just part of the old mill.

The main shop had two doors, wide enough to admit an automobile if the interior was cleared out. Outside the doors was a small parking area on the uphill side of the building. Opposite the doors, windows looked over the millstream. The front of the mill faced onto the road and the millpond, which barely could be made out through dust-covered old window glass. To the rear, the shop led through a doorway to a back storeroom that contained even more shelves piled helter-skelter with hardware. I had spent many an hour sorting through these items that had gathered dust slowly over two generations. There were familiar objects: brass screws of many sizes in small containers, drill bits. And there were many pieces of hardware deposited there from an old blacksmith's shop that had once stood upstream. The shop had been taken down when it threatened to fall down of its own accord. These objects, having to do with horseshoeing, the construction of horse-drawn wagons, coaches, and sleighs, were not comprehensible to me. I believed that few people had looked at some of these in half a century. Sometimes when I stood there, I imagined that Heman's stepfather, long dead, was watching over my shoulder. I wondered if he would be pleased to see a stranger such as myself fascinated by what he had casually put aside one day and then forgot about, item by item. Sometimes, standing there, I had the feeling that he was a presence become palpable, and I had to leave. So it is with dusty corners of old buildings filled with large pieces of a technology no longer quite understood. I wondered if someday the radiated forest at Brookhaven, then so modern and advanced, would be another such attic of devices and memories. Both the new nuclear laboratory and the old mill held concentrations of machine parts, but these collections were as different as they could be, the one up-to-date and orderly, the other whatever a person had been using and just set down wherever it was handy. The way bits and pieces of technology were stored seemed to epitomize the two views of life, although I couldn't at that time quite put my finger on what that difference was.

Repairing the mill brought me in direct contact with a way of life different as could be from the hi-tech science at Brookhaven National Laboratory. Here was a small, industrious New England village, almost self-sufficient except for the need to import certain raw materials such as the metals to be worked on in the mill, or some complex tools made outside. In exchange, it produced trade products from the series of mills that once filled this stream valley, as well as milk from local dairies and other agricultural products, and some timber cut from local forests. Most of all, it brought me into contact with a now-vanished world view, a world in which how things worked was plain to see, bigger than a breadbox, explainable to a casual passerby.

The mill was as near a place of certainty in one's ability to control nature's power and sustain the use of that power as any place or any system I had known. It was part of a past world in which a small village of similarly minded people could lead lives that seemed reliable, dependable, and determined by one's own actions. Even in the cold winter, with ice on the flume's water, the mill, lively with good conversation, hard work, and a seeming certainty that we would achieve our goal, seemed as stable and predictable a world as I could imagine. The intense ambitions that at Brookhaven and other major scientific laboratories sometimes led one scientist to denigrate and try to destroy the reputation of another had no presence here.

Heman loved his surroundings—the streams and the woods. He loved his work and he loved the people of New England. His idea of nature was a place to live within and enjoy, and a source of resources that one husbanded carefully. Researchers at Brookhaven's radiated forest enjoyed the woods, but it was more a thing external for one to study than a way to live within.

Heman was careful in his treatment of the mechanical equipment, but he took electrical equipment, that more modern, less visible kind of technology, casually. The wiring in the mill was the old-fashioned kind, made of separated copper wires covered by cloth insulation; and these, running in parallel like toy railroad tracks, were attached to a series of wooden laths. The pegs on the laths holding the wires made convenient places to hang things, and many of them had clothes hanging from them on metal hangers, with the hangers touching both electrical wires. If the old cloth insulation wore through, the result might be quite spectacular sparks and perhaps a fire in the wooden mill building.

Heman had a big power saw that he had made himself from an old washing machine motor, a wooden housing he had built, and various old pieces of hardware from the mill's shop. To turn it on and off you had to use a big knife switch, a kind you don't see today—a U-shaped copper bar about four inches wide, hinged at the open ends and with an insulated rubber handle on the bottom of the U. To turn the saw on, you had to grasp the switch handle—as wide as your hand—and push it into a pair of copper clips. This completed the electrical circuit and brought electricity to the motor, which responded immediately with a loud whirr to spin the saw blade. Heman had mounted this switch on the right side of the saw, about knee high. To turn on the motor you had to bend over the saw and its big exposed circular blade, take hold of the rubber handle, and push it upward so that it fit into copper grooves that connected to the wiring of the motor. From my point of view, turning on this saw carried a double risk: If I touched the copper bar instead of the insulated handle, I might get a mean shock; and leaning over the saw, I might cut off my head or cut deeply into my chest if I slipped and fell forward onto the blade as it started.

After a few years I couldn't stand it anymore and I rewired the entire mill myself with then modern plastic-enclosed wiring and put a fancy, sealed, heat-protected on/off switch in a convenient location in front of the power saw. Heman looked at the elegant switch and, in typical New Hampshire fashion, said nothing but gave me a puzzled grin, as if to say, "Nice of you to buy that Cadillac of a switch for this homemade saw and I'll use it, but I really never needed it." The rest of us, Heman's wife, Edith, especially, were happy with the new safety switch. It seemed to me that Heman drew a blank with the new electrical and electronic technology, not taking it seriously as dangerous, nor considering it something to be respected and made beautiful, in contrast to his respect and love for old brass and chrome-steel machines.

After several more cold days of work, Heman and I finished installing the new flume, and the mill sprang back to life, vibrating, spitting out wood chips, spewing water through the turbine wheel. Reliable machinery of good iron, steel, and brass, carefully made, artfully designed.

"This is the way everyone should live," Heman said to me as we watched the mill work once again. "Be independent. Nothing better than a country village life. Trade as little as possible with the rest of the world. Do what you want. Rely on yourself. Make and fix whatever you need." The clear implication was to keep clear of big government, big projects, and big science. The radioactive forest with its complex and, for the time, miniaturized computer equipment, its mysterious silent radioactive source, its loud pumps and paper punch, its miles of electrical wires and plastic tubes, seemed another world, another universe. Yes, I seemed to be living in two centuries, the nineteenth and the late twentieth, each unexplainable to the other. I did not want to have to choose between them and believed that science and engineering could carry me through to enjoy them both indefinitely.

Somehow, I had to put the two worlds together—the mill and the streams and forests in New Hampshire, and the computers and radioactive forest on Long Island. I was trying to find my place in nature and in civilization, to understand how nature worked and how societies functioned. The difficult, strange, poignant, and funny incidents that this attempt would cause me were still in the future as I helped Heman attach the final section of the flume to the turbine waterwheel, our feet standing on frozen stream water in the midst of a New Hampshire winter, now so long ago and so far away.

Four

LOST IN AN
AFRICAN WILDERNESS

 few years after I worked in the irradiated forest and spent time in New Hampshire helping Heman Chase fix up his mill, I began to travel to East Africa to study elephants. On these trips I got to know Roger Atwood, another American, who had moved to Africa and become one of the world's experts on the big game animals. Through his hospitality, I heard something that a long-term resident in Kenya told me about wilderness that has stuck in my mind ever since. Roger had married into a well-established family, originally from France and Italy. During one of my trips to Kenya to study elephants, he was a kind and highly informative host, flying me about in his Cessna 182 over Masai Mara, the famous park at the north end of the Serengeti plains, and over the main Serengeti plains themselves in Tanzania. As we bounced in the turbulent air just above the ground in Roger's airplane, one day when we were going out to one of the parks to study wildlife, I realized that here again as at Brookhaven National Laboratory and in the old New Hampshire mill, technology and nature had some deep connections.

Near the end of our trip, Roger invited me to the home of his mother-in-law, Angelica Le Fabre, on the shores of Lake Naivasha, a famous and beautiful lake north of Nairobi where many of the great white hunters of earlier days in the twentieth century built homes. The home was a magnificent compound with well-watered lawns and trees, houses for the staff, a large and comfortable guest house, and the main house, which fronted onto the lake and was a handsome art deco example from an earlier period in the twentieth century.

After a delightful dinner, we sat outside on the lawn in the long East African evening and Roger's mother-in-law talked about her life in Kenya. She was born in France and came to Africa with her husband well before World War II, in the 1920s, she told us. "When we first arrived, there was nothing here," Angelica said, "nothing."

An African dove sang incessantly in the trees, its song the constant reminder of the wonderful, open feel of the African landscape and its sense of wildness.

"Do you know that for the first two years we just wandered in the wilderness, never knowing where we were?" she said, as the darkness gathered around us.

"Really?" I responded, genuinely impressed. I looked around in the fading light at acacia trees and thought about the vast savannas and their abundant, beautiful, but sometimes dangerous wildlife that I had just spent weeks watching. I thought about the book called *The Last Best Place*, which described the Serengeti plains to our south as the last perfect wilderness on Earth. I began to imagine myself in that landscape in the 1920s before modern roads, with few European-style houses, lost among the lions, leopards, elephants, and the black and white rhinoceros, trying to find my way from place to place, trying to avoid contact with the most dangerous of these animals, trying to survive. Coming out of my brief reverie, I asked, "My goodness, what did you do for food?"

"Oh," she said, "we had fifty porters."

And so it is. Each of us has his own idea of wilderness, influenced by our background, culture, and recent experiences. She and her husband may have wandered for two years in their wilderness, but they were going from village to village—villages of the porters' families—in a well-lived-in, highly humanized African countryside.

While I sat in the pleasant Kenyan villa, thinking of one person's idea of wilderness, I thought about another reaction to nature, far away in New Hampshire, many years before. One year I lived in a New Hampshire colonial farmhouse, a small, weathered pine-board-sided farmhouse. My nearest neighbors were Elsie and Clarence Goodenow, a sister and brother who had never married. They lived their entire lives in the house where they were born, in Acworth, New Hampshire, running a small dairy farm. Acworth was a small village about fifteen miles east of the Connecticut River, between Brattleboro, Vermont, and Keene, New Hampshire, and about six miles toward the Connecticut River from Chase's mill in Alstead. Unmarried siblings who spent their lives together were not so unusual in New England. Elsie occasionally helped out at the Chases' with cooking and cleaning, to bring in a little cash, and she had become good friends with Heman and his wife, Edith, and their two daughters, Ellen and Margaret.

The Goodenows' farm occupied a stretch of land on upland slopes facing south, just above a small river. A mile or so downriver was an old covered wooden bridge, still in use at that time, now bypassed but intact. They had about fifty cows, and their land was open pasture mixed with strands of northern hardwoods and conifers—sugar maple, yellow birch, beech, white pine, and, on the higher slopes, a little balsam fir, red spruce, and white birch. The cows walked through woodlands on their way between pastures. The entire landscape had been glaciated about ten thousand years ago, before the valleys were steep and U-shaped. The riverbed was mostly gravel, and the water flowed clear between the steep slopes.

Elsie and Clarence were cheerful and friendly and had that wry sense of humor and propensity to say absolutely no more than was required, no extraneous words, for which northern New Englanders are famous. We visited the Goodenows on occasion, always enjoying their company. Their house was an old New England–style clapboard, of some no longer distinguishable color, badly in need of painting. There was a covered front porch, rarely used, an old car and old truck or two parked on the grass near the house, and various pieces of farm equipment scattered here and there. The inside was small but tidy and we sat around the living room's woodstove, a 1930s model on which Elsie set the coffeepot to keep it warm.

One January Sunday morning, we were visiting and drinking coffee. As usual for that time of year, snow covered the ground. But the night before, there had been a warming, and freezing rain fell mixed with snow. This morning the sky was clear, the air cold, and all the trees were covered with ice down to the tips of their branches and twigs. On the way over, we noticed how beautiful everything looked, with the trees shining in the sun. Some of the iced-over branches were acting like prisms and reflecting bright rainbow colors.

An hour or so into our visit, Clarence said it was time he checked the cows, and went off to the barn. He returned a while later and Elsie asked, "What's it like out there, Clarence?"

"It's a goddamned fairyland," Clarence replied.

The Goodenows were part and parcel of the nature they knew. They never sought to experience wild nature, but they loved the farm and their cattle and had fun with life.

My thoughts returned to the Kenya villa. I listened to the incessant calling of the African dove. Many of us romanticize about wilderness, although few of us have the opportunity to experience firsthand the wilderness of television's nature programs. Fewer have the fortune to appreciate it as directly as Elsie and Clarence Goodenow. Most of us are urbanized or suburbanized creatures, whose knowledge of wilderness is vicarious. We are as isolated from real wilderness as was Roger At-wood's mother-in-law. Her impressions of Africa illustrated how impor-tant preconceived, culture-specific beliefs can be to our observations. When it comes time to discuss the true character of nature undisturbed by human influence, we should keep the African mother-in-law and her fifty porters in mind—and the goddamned fairyland of a New Hampshire dairy farm.

Five

❧

AVOIDING DEER IN THE FOG

ne of the problems with trying to understand nature is that usually chance or some kind of randomness affects what happens, so you cannot predict things exactly. It's like the weather. Weather forecasters used to say things like "rain tomorrow" or "sunny all day," but now they rephrase their forecasts as the chance of rain or sunshine—"forty percent chance of rain." I soon became aware of this kind of limit to what we could know about nature. Joe Pitalka, a friend of mine, worked for the New Jersey Department of Fish and Game and knew a lot about deer—their behavior, their food habitats, how to manage them. One day he was riding with his friend Mike in Mike's car along the coast of New Jersey near Atlantic City. They drove into a thick fog where they could see little in any direction. Suddenly, a deer dashed across the road in front of them, a quick ghostly movement barely missing the front grille.

"Better slow down," Joe said to Mike. "Where there's one deer, there's usually others."

Mike slowed down. A few minutes later, a deer dashed out of the fog and ran into the side of the car, killing itself and doing severe damage to the door. If the car had not slowed down, the deer would have passed safely behind it. Joe understood deer well and his advice to Mike was based on that sound understanding. But understanding nature never leads to a perfect forecast. He was quite right about the group behavior of these animals. Deer often move in small groups and will cross a road together. But when the deer reach the edge, each hesitates a different length of time. Some are shy; some are bold. You can't predict exactly, from a speeding car, when each deer will cross a road. Ironically, if there had been nobody in the car who understood deer, everything would have been all right.

So it is with nature and life in general. Sometimes no amount of knowledge is enough to make us completely safe or certain. There are always some risk, chance, and uncertainty, so that even the best forecasting methods can go astray.

In thirty years of studying nature and trying to understand it, I have found that, just like dealing with deer running in small groups, we can make a reasonable forecast that a certain event will happen, but we are not very good at predicting when or where. Scientists call this the difference between a qualitative and a quantitative forecast.

This kind of uncertainty is a problem we have with all of nature and all of life. In the United States we seem to want to legislate and regulate such uncertainties out of existence. I became keenly aware of this when I moved to Santa Barbara, California, in 1979, long after I had helped Heman Chase fix the mill in New Hampshire and I had carried a Geiger counter over ground zero in the irradiated forest—about the same time that I sat by Lake Naivasha and heard about living in an African wilderness. Driving in the countryside near elegant Santa Barbara, with its golden brown grassy slopes reminiscent of the African plains, its steep chaparral-covered dark green mountains, and its bright blue ocean, I was looking to buy a house. I was surprised to see houses built on steep slopes, on slopes of soft soils that turned to moving mud in heavy rains, and next to dry stream channels, right down on the floodplain. We first rented a house in a pretty but narrow canyon just to the west of town in an upscale housing development—the kind with a stone arch with the name of the development in large letters at an entrance to wind-ing roads. The creek bed, dry at the time we moved there in August, had eroded a two-level stream channel. The channel encompassed an upper floodplain where sediment eroded from the hills had been deposited fairly uniformly in the valley as the stream had meandered back and forth over the centuries, and a lower, present-day stream channel, cut below the larger floodplain. The present channel was perhaps as wide as two houses. Anyone observant about natural history could see that the stream filled the entire channel during floods, and floods had to occur now and again, but there were some houses built right down next to the dry streambed, disasters just waiting to happen. And when such disas-ters happen, we seem to have come to assume that it is our right to live there and that somebody owes us payment for what we have lost. We do not want to accept the risk and uncertainty about nature's climate. The real world of chance and uncertainty is hidden behind clouds of assumed certainty; this leads to misunderstanding and misin-formation.

When we did find a house we could afford—above floodplains and away from the worst wildfire dangers—we had to decide what kinds of homeowner's insurance to buy.

I asked Tom Lyon, a geology professor at the University of California in Santa Barbara, whose field of research was earthquakes, if I should buy earthquake insurance.

"Well, I don't have it," he said. "The deductible's ten thousand dol-lars. If an earthquake did ten thousand dollars' damage on average to each house in town, it would bankrupt the insurance companies and the federal government would have to come in and bail everybody out, so in a sense, we're all insured anyway for free. The insurance premium is high and the chances of the kind of earthquake that the insurance would cover are very low. Now, if I lived in a house that was mostly glass and did not have the kind of wooden framing that gives when the earth moves, or if I lived in a house on a landfill, I might buy it," he added. "Of course, you've got to make your own decision." I decided not to buy earthquake insurance.

But I did buy insurance against wildfires, as did everybody else. Wildfires were common, the deductible was low, the premium was reasonable, and if you did not have wildfire insurance you would be out the value of the house, something I could not afford, nor could many people.

The choice of what kind of insurance to buy in each case, earthquake and wildfire, was rational—based on an acceptance of the natural uncertainties and risks about nature. This is the way we need to deal with environmental problems in general.

I have done some research related to global warming—looking at the possible effects of global warming on forests—and people continually ask me if I believe that global warming is "going to happen." Then they ask, "If so, when, and should we do anything about it?" I tell them that the scientific community overwhelmingly accepts the idea that our burning of fossil fuel is leading to global warming, but that there is uncertainty about the amount of that warming, the effects, and the timing of those effects in any location. I suggest that they think about global warming as a risk in the same way that you have to think about earthquakes and wildfire in Southern California—in terms of whether we, as a society, should buy the equivalent of insurance. The way to decide is to consider the probability that global warming will happen and have costly effects. We could engage in a set of actions to minimize the rate of global warming as a sort of insurance policy. These actions would include planting trees and reducing our use of fossil fuels—turning to wind,

solar, and geothermal energy. These actions are beneficial for many reasons: trees are useful and beautiful; some tree plantations actually make
money; trees provide habitat for wildlife—including many threatened
and endangered species—as well as recreational opportunities, shade in
cities, and shade to cool our houses. Even if global warming never happened, tree planting on a large scale would be beneficial. Wind and solar
energy are plentiful. The technology available today is reliable and can
be obtained for reasonable cost, especially if government subsidies for
fossil and nuclear power were removed or reduced. So the "cost" of this
set of actions—the equivalent of the cost of an insurance policy against
global warming—are low or actually a positive gain.

Next we need to consider how damaging the effects of global warming could be if it were to occur. There is much controversy about this,
but the key is how global warming will affect lives, human and otherwise.
The effects could be huge. How huge? There are four or five different
computer models of climate change, and their forecasts vary in how
much the change will be. All of them forecast an increase in temperature.
One forecasts an increase in rainfall in one region of the Earth, while another forecasts a decrease. But in trying to analyze possible effects on
forests, I noticed that the forecast temperature increase was always large
enough to dry out the soil more than rainfall would add moisture to it.
When global warming occurs, the land will become warmer and drier.
Water for irrigating crops will have to increase, and we are already using our water supplies faster than they are being replenished. Forests
will suffer from drier soils as well. We will have less water for our
homes, gardens, lawns, and parks. Sea level will rise, and large areas in
low-lying countries could be flooded. Some forecasts suggest that nations like Bangladesh could lose as much as a half of their agricultural
land. So the effects are severe. Considering the equivalent of the costs of
the premium and the effects of global warming, the same way I considered whether to buy earthquake or fire insurance for our Santa Barbara

house, the conclusion seems clear enough—do what we can to reduce the probability and rate of global warming.

Chance is everywhere, in the possibility that you might drop your radiation badge in the midst of an irradiated forest, that you might cut yourself bending over an old power saw in an old mill. It was in the turbulent air over the Serengeti in Roger Atwood's airplane and whether or not he would be attacked or ignored by the elephants he studied. Yes, it might even be in the chance that you would have to work out the value of Whorehouse Jack's business in Idaho. So when you are faced with discussions or decisions about nature, or for that matter life in general, remember the deer that crossed the road in the New Jersey fog and the man who knew about deer and did his best to avoid them. Plan for probabilities; survive with uncertainty.

Six

❧

THE MONKEY'S DILEMMA

e were standing in a rain forest of Costa Rica, a light rain falling around us, our shoes caked with mud. It was near dusk, and within the dense layers of leaves the forest was dark. We didn't know what was wetting our clothes more—rain or sweat—as we stood and listened to Ben Sheldon talk to us about a strangely but neatly cut leaf of a palm tree. We had hiked for several hours and were not sure when we would reach our destination— Ben's eco-resort, Paradisio Perdu—nor the distance it would take us to reach it. He was telling us a peculiar story about bats, monkeys, and palm leaves.

I had come to Costa Rica to work on a project in its rain forests, and several of us were on a weekend field trip to see what we had been told was one of the most interesting and beautiful parts of these forests. Ben had set up the resort to introduce foreign tourists to tropical rain forests and to show that this way of conserving the forests could pay for itself.

It was a private, rather than a public, governmental, method to help save nature.

Four of us made the trip: Mike Marzolla, a friend who had come to Costa Rica as our translator and guide; Lloyd Simpson, a postdoctorate working for me in a study of the rain forests; and Lloyd's wife, Kathy. We had driven for several hours from San José, the capital city of Costa Rica, to a turnoff where we parked at a small house and farm. A sign told us this was the starting point for the ride to Paradisio Perdu. A pretty little girl in a muddy dress stared at us from the door of the house. We and the other tourists got into a long cart mounted on big wheels, with a canvas roof supported by wooden slats. This kind of cart was used to pull farmworkers around Costa Rican banana plantations through mud and across rivers. A red, American-made four-wheel-drive tractor with a crew of three Costa Ricans pulled the cart. The cart filled up with about a dozen people who were going to visit the private nature preserve.

Soon after we started off, the cart and tractor crossed two rivers—rocky-bottomed, clear-flowing rivers, lined on both sides by rain forest and farms. Water came almost up to the floorboards and the cart seemed at some points about to flow off into white-water rapids that we could see just downstream. But the tractor huffed and pulled and got us across the rivers and onto a muddy track that passed through cattle ranches, green fields, and scattered patches of intensely brilliant green rain forest trees.

The ride in the banana cart took three hours over a very bad road composed of a heavily eroded red soil, almost a clay. At times, the tractor and cart got stuck among the mud and boulders, so we had to get out and walk while the crew pushed and pulled and manhandled the vehicles. The stops got more and more frequent; we seemed to be walking as much as riding. It was a pleasant day and the walk would have been nice except that the claylike mud of the track stuck to our shoes and made walking slippery.

We made a lunch stop part of the way up. There Ben met us, having walked down from his resort. He invited us to walk the rest of the way, rather than ride, and he would tell us about the rain forest. "It's only three kilometers, an easy hike," he said. This seemed a pleasant invitation and we agreed. He told us that the resort was relatively inaccessible on this terrible track because he wanted it that way, so that logging trucks could not use his roads to get in and begin to cut down the big trees.

Our walk soon became quite strange; risk and uncertainty were afoot. Ben walked us through water that was halfway up our boots, making the heavy clay soil and rocks even more slippery. In places, Ben had made a rough trail out of unfinished, half-cut boards that tipped on rocks beneath them. Any of us could have fallen and broken something, especially at the pace he kept.

As we walked, Ben told us about the local animals and plants. Now and again he stopped and stood in a drizzle under the trees. He was dressed in a T-shirt and old gray trousers that had been resewn many times, with rubber boots on his feet. He talked rapidly and looked a little wild-eyed and intense. By mistake, he had zipped his fly so that the tail of his blue T-shirt stuck out at the top of the fly, like an oxygen-starved penis.

He had been born in Brooklyn, Ben told us, but had lived in Costa Rica for more than twenty years. He was like other expatriates, American and European, that I have met in the tropics, especially on the outskirts of civilization—the sort of person you run into when you least expect it, although you get to learn that such people do live there. As with the deer crossing the road, you can feel certain that they are there; you just don't know when they will turn up. They're like characters in a Joseph Conrad novel or the photographer who turns up in the midst of the Southeast Asian rain forest in the movie about the Vietnam War, *Apocalypse Now*—people who, for one reason or another, do not fit in at home and have left their native country to make themselves important in a developing nation or to act out behaviors not acceptable in their home country. They are often interesting, but there is also something a little strange about them. They are another one of the surprises and bits of chance that you meet when you are out to try to study nature in a re-mote area.

Mike told me later that he noticed there were two trails that led from the tractor track and around to the residential part of the resort. Ben had taken us on the longer one. He had not given us a choice, nor asked us how we felt—if we were tired or hot. It took us about three hours to go three kilometers, or what he said was three kilometers. When we had hiked for two hours, I finally said to Ben, "Where is the lodge from here?" Ben pointed up the trail and said, "That way," which tipped me off that he was really testing us. I never asked again, but he did offer at one point that it was about a kilometer farther on. He had gone out of his way to create additional uncertainty for us. Mike later referred to this as "the forced march." It was beginning to get toward twilight, but Ben had one more story to tell, which turned out to be worth the entire trip.

He took us off the trail a short distance to a small palm tree that, he told us, grew only in the understory beneath the tall trees of the forest. A large frond of the palm, perhaps three feet long, had been cut on both sides, sharply as if with a knife, from the edge to the large vein in the cen-ter, which had been left intact. The intact frond had a natural angle to it, and the cut made the outer end of the frond into a kind of little four-sided canopy. "Bats of a certain species make this cut," Ben said. "They like to build their nests under the frond; it's protected from the rain. They use the nest for only a few years. Then they abandon it and make another somewhere else."

"Monkeys like to eat these bats," he continued. "But once the bats abandon a nest, forest wasps often come along and use the nest as a base to build their own. The bats are fast, and if a monkey peers under the palm leaf to see if a bat is there, by the time he looks and grabs, the bat is gone. But if a monkey grabs without looking, he might get a bat, he might get nothing, or he might get a handful of stinging wasps."

This risk was inherent in the monkeys' lives. There was no way to know exactly what was under any specific leaf of this palm. The monkey who wanted to eat a bat had to take the risk or go away hungry. A monkey came to learn that some of the cut palm leaves did have bats and some had wasps—that much was certain. This was the monkey's dilemma—whether to grab under a leaf or not. It was just like my friend who was an expert on deer in New Jersey—he knew that deer often went in small groups. But in neither case could either the monkeys or the deer expert know exactly how any specific situation would turn out. It's a part of nature that is difficult to understand and makes it impossible to offer predictions with complete certainty. Chance is part of Mother Nature's kit of tools to make studying her difficult. It places a limit on what we can know.

Meeting Ben Sheldon was a little bit like this as well. It seemed a roll of the dice when and if we would meet this kind of expatriate character on this trip to Costa Rica—I hadn't run across any others in that country on my previous trips, but, as I said, you can be pretty certain that his sort of character is somewhere in the outposts of civilization.

We had the opportunity to learn firsthand more about risk and uncertainty at Paradisio Perdu the next day. We stayed overnight in a pleasant wooden building, octagonal-shaped, where every room had a patio with a view of the nearby rain forest or down an open, grassy slope, to mountains in the distance. Mike and I shared a room and enjoyed sitting outside and watching the beautiful tropical birds. The only obstacle to our view of the distant mountains was a large and ugly cabin that served as kitchen, dining area, shop, and headquarters of Ben's resort. It seemed odd that he had chosen to place the headquarters building directly in between the most beautiful view and the residential building, a strange design by an expatriate hidden in the outposts of civilization.

The next morning, we walked over loose and rocking boards that were supposed to keep our feet above water that ran down the muddy slope in the clearing. Slowly we made our way down to the headquarters cabin, where we breakfasted on the porch. Viewing it up close, we could see that the cabin was crudely put together. Its primary decoration was a large skin of a huge snake that Ben had hung on the wall. The porch was on the upslope side of the building, so the beautiful view of the moun- tains and forests was cut off from us. We ate looking at the snakeskin.

After breakfast, most of the other visitors, led by Marjorie, Ben's as- sistant manager, a British woman with a funny Americanized accent that sounded Australian, went on a tour of the rain forest. Mike and I de- cided that we had learned a lot about Ben's forest from him the day before, and we wanted to visit a famous waterfall on his property and to watch the birds, best seen in openings rather than within the forest. After the others left on their hike, Mike and I walked down to the falls. The path went through wetlands where Ben had once again casually placed some loose boards that shifted under our feet. It was easier, but more risky, than wading through the claylike mud. At the streamside, there was a crude path down a steep and muddy incline, where we al- most slipped several times.

We found a small stream, maybe twenty feet wide, that flowed over a short falls and then over a very long and beautiful falls. We walked down below the second falls, admired it, took photographs, and talked with Ben for a while. Ben told us that visitors swam in the pools below the falls. Looking down, we saw that the water rushed quickly and ap- peared dangerous to all but the best swimmers. Since most of the other visitors seem unprepared in any way for a trip into a wilderness, we thought that a swim below the falls could pose a risk to that kind of vis- itor. The slippery paths also made us wonder how an average tourist without any standard hiking equipment might make out in this isolated place. We were soon to find out.

About eleven o'clock Marjorie returned from the hike alone and said very quickly that one of the visitors, a retired American schoolteacher, had fallen and broken her arm. The rest of the party eventually arrived, helping the lady with the broken arm. She lay down on the porch of the headquarters building, beneath the snakeskin, and was clearly in considerable pain.

Ben had no splints of any kind in preparation for such an accident. He had no stretcher and almost no first-aid equipment. He seemed totally unprepared for this or any contingency. Although he had told us a great story about natural risk and uncertainty, ironically he seemed not to have learned the lesson his own story implied: expect the unexpected; prepare for probabilities. Paradisio Perdu had no painkillers, not even aspirin, and the schoolteacher was in great pain.

Ben went off and got one of his workmen to cut some lengths of lumber he had lying around—1-by-2s—to begin to make reinforcement for a kind of splint. He came onto the dining porch carrying these and a big piece of plastic foam and proceeded to whittle and slash at the wood. I thought that this seemed to be incredibly unprofessional and dangerous. I decided that it was time to take some action to help the poor lady.

Lloyd, who was in charge of field crews that worked in remote forests all over North America and Costa Rica, had had more experience than I in emergency situations, and his wife, Kathy, was trained in first aid. I talked to Lloyd when he was packing his suitcase and suggested that he come down and help out. Lloyd and Kathy came down immediately and did a magnificent job, taking over completely from Ben, rapidly making a good splint and gently putting it on the woman's arm.

We asked Ben and Marjorie if they could call in a helicopter to evacuate the lady with the broken arm; we dreaded the thought of her riding for more than three hours in the bumping banana cart. Marjorie asked if the lady had the two thousand dollars for a helicopter, saying there were only four in Costa Rica. Later, Mike overheard Marjorie say it would taken eighteen hours to get a helicopter and she didn't want that complaining lady around for eighteen hours. Kathy, distraught, asked if Ben had insurance that would pay for a helicopter flight. We were all shocked to discover that there was no equipment, no preparation, no insurance, no plan for dealing with this kind of contingency in a place that was an accident waiting to happen.

By the time lunch was over, Lloyd and Kathy had finished setting the lady's splint, and she was carefully walked down to the banana cart with Lloyd, Kathy, and several of Ben's workmen holding her and guiding her as tears streamed down her face. Kathy, Lloyd, and Mike suggested that perhaps she could be carried out in a kind of stretcher rather than driven in the cart. Mike, who was fluent in Spanish, spoke with the Costa Rican staff and they offered to carry her out this way. They added they would prefer to carry her all the way out because this would be much better for her than being bounced around in the bumpy car on the bad road. But Ben insisted that she go on the cart.

The kindly Costa Rican driver of the cart went very slowly because the bumps were so painful to the schoolteacher. We left Paradisio Perdu at 2:30 in the afternoon and reached the little village around ten o'clock at night. What had been a three-hour-plus ride up became a seven-and-a-half-hour ride down. All the time, the lady with the broken arm was in terrible pain. Lloyd sat next to her, supporting her continually. Mike sat next to Lloyd, putting pressure on him so he did not get thrown around by the rough ride and could steady the lady. Kathy kept the schoolteacher's arm supported by a big foam cushion she had made up from material scavenged around the resort, and also supported the artificial splint. The lady was given a few grams of codeine that somebody had in their pack to ease her pain.

Kathy was furious with the amateur nature, lack of preparation, and poor condition of Paradisio Perdu. The woman's companion, a nice person from Dallas, came to the resort only in shorts and a thin tank top. As we rode down, it began to rain and the water blew onto us under the canopy. The lady with the broken arm began to look poorly. She reminded me of people who get hypothermia in the cold mountains. I asked her if she was warm enough as we came down through the mist and the rain. She said something quietly to the effect of "not really" so I lent her my sweater, something I always carried in a backpack in back country, no matter what the climate, just for unlikely chance occur-rences. Marjorie walked behind the cart, having left the resort without so much as an umbrella. She ended up borrowing my umbrella and re-mained unhelpful and outspokenly unsympathetic the entire way down.

We crossed the river in the dark, once again with the sensation that we would be swept away into the rapids that we could only hear, not see, in the dark. Finally, we reached the parking lot. A radiophone call had requested an ambulance, and it was there waiting for us at the end of the trail. The injured woman was taken to a hospital. We learned the next day that she was doing well and had received good care at the hos-pital in the capital city. I wondered what might have happened to her but for the knowledge and kindness of Lloyd and Kathy. Chance had caught up with her, just as it did with the monkeys who grabbed under a palm leaf and came up with a handful of stinging wasps. She had learned firsthand, a very hard way, about risk and uncertainty in nature and also about the kind of resorts built on big ideas but small budgets in the wilds.

A few years later, we read in a newspaper that two German tourists had drowned at Paradisio Perdu while trying to swim below the falls that Mike and I had visited. It seemed that not much had changed.

We often think that we can solve problems completely with total foreknowledge, both at home and on vacation. Some of us wander into the wilderness with little preparation, assuming that somebody will al-ways bail us out of danger or, if not, the government is obliged to and will. But our trip to Paradisio Perdu always serves as a reminder to me that this is not true. We live in a world of chance. In that world, no so-lution can be perfect, no forecast can be made without some chance of error, no trip to visit wild nature can be completely safe. Sometimes we

can estimate the probability of what we wish would not happen and prepare for those contingencies. That preparation was a matter of course
for our field crews, but was not typical for the average tourist to Costa
Rica, who had set out to enjoy a wonderful experience to see a tropical
rain forest at Paradisio Perdu, as several tourist guides described it, making it sound perfectly and completely safe. Sometimes we are lucky and
all goes well. Sometimes you can break your arm, as did the nice lady at
Paradisio Perdu. Sometimes even when you do this, you can be lucky
once again and there will be people like Kathy and Lloyd nearby to help.
Sometimes you will be stuck entirely on your own. We can never know
exactly when and if we will grab a solution or get badly stung. Life, it
seems, is a series of events that are best thought of as, well, the monkey's
dilemma.

❧

Seven

HOW MANY HOURS
DOES A WHALE SLEEP?

ow many hours does a whale sleep?" I asked into the telephone receiver. If anybody could have seen my red face, he would have known how embarrassed I felt asking this question. I was working as a research scientist at the Marine Biological Laboratory in Woods Hole, Massachusetts, doing a professional, scientific study of whales. Here I was on the phone to the New York Aquarium, asking the kind of question a child might ask of a zookeeper.

"You want to know what?" a voice said at the other end of the line.

It was about the tenth—I had lost count—person I had called that day. Each person who answered had reacted the same way and had shuffled me on to someone else.

"How many hours does a whale sleep?" I said.

There was silence at the other end of the line.

"Why do you want to know that?"

"Well, it's kind of a long story. I'm doing a study of whales . . . I'm at MBL," I said, hoping that the well-known nickname of the famous marine laboratory might sound like a justification.

"Do you have any whales?" I asked.

"Beluga," the voice said. "One beluga."

"How long does it sleep?" I asked.

"Hold on a minute."

I had been calling around the country for the whole day, as had my research assistant, asking the same question. It was getting late in the afternoon and I was tired and frustrated.

"Hello," said the voice. "We're not really sure. Maybe two hours. We don't watch the whales to see when they sleep. We've more important things to do. It's very busy here with tourists and then caring for the fish. Why do you want to know?"

"I'm a scientist at the Marine Biological Laboratory," I said again. "We're doing a study of whales—whale behavior. It has to do with the conservation of whales."

I didn't know if my question sounded more foolish to an aquarium expert coming from a scientist or from an eight-year-old.

"It has to do with the International Whaling Commission," I said, hoping this would impress the voice.

"Why do they want to know?"

"The Japanese have asked permission to take some of the big male sperm whales," I said. I had gotten the voice's attention, I hoped.

"Why do the Japanese want to know?"

"They don't. I do. A conservation organization asked me—us—to find out what the harvest of these big males might do to the behavior of groups of sperm whales—and then how it might affect the number of calves born," I said.

"Well, if you say so," said the voice. "Look, I've got to go. One beluga whale. Two hours. Maybe."

I looked out the window of my laboratory office to the "Hole," the narrow passage between the mainland and Martha's Vineyard, where ferries went to and from that island and Nantucket. It had been a long day and I was beginning to wonder why I continued the struggle to get an answer to what seemed an absurdly simple question: How long does a whale sleep? Other phone calls had yielded precious little. Another aquarium keeper told us that one of their killer whales seemed to sleep very little. One of the whale scientists at Woods Hole told me a story about a boat that had banged into a sperm whale that seemed to be dozing. The whale had been startled by the collision. "But what does it mean for a whale to 'sleep'?" he had asked, then answered, "Perhaps whales rest but don't sleep like people." This was a philosophical point I wasn't able to deal with. What did I mean by a whale "sleeping"? Is it a specific state of its nervous system? Did a whale dream? I decided that all we needed to settle for was a condition like the sperm whale, plain and simple, lying on top of the water, seemingly inactive and unaware of its surroundings until "woken up." We had to throw out the philosophical or neurophysiological question of what "sleep" means in general and what it might mean for a whale. We weren't about to put radio collars on whales with detectors for brain wave activity.

I had just recently taken the job at the laboratory at Woods Hole. The town and countryside appealed to me; they seemed a neutral ground between nineteenth-century, rural New Hampshire and twentieth-century, hi-tech Brookhaven National Laboratory. Woods Hole was a small village, part of the town of Falmouth, Massachusetts. In the nineteenth century, scientists and naturalists interested in the ocean and its life had come to the Hole, and in 1886, some had set up the Marine Biological Laboratory. It was right on the waters of protected bays and inlets, and seawater was pumped directly into the laboratory so that research with sea life could be done easily. Summer courses were famous there. My sister, Dorothy, had taken one when she was in college. The laboratory had a beautiful, modern brick residential building with a central cafeteria that looked out onto the Eel Pond, a small, squarish salt-water pond that served as a marina, connected to the Atlantic Ocean through a small inlet over which stood a tiny drawbridge.

I stared at the ocean and thought about how this had all begun. A few months before, Charley Russell, a Canadian mathematical economist who sometimes worked on conservation of whales, turned up unannounced at my office door.

"Hi," Charley said. "I was in Woods Hole and wanted to talk with you about a project. Maybe you'd be interested. I was just at the annual meeting of the IWC." The IWC is the International Whaling Commission. It met every June in Cambridge, England, and Charley was on its scientific advisory board. Charley was thin and wiry, and was dressed, as usual, in a corduroy sports jacket and a dress shirt unbuttoned almost to his belly button.

"Christ, what an ugly place," he said with a sweep of his hand.

"Sorry, my office is messy," I replied.

"No, not your office. The women here in Woods Hole," he said. "They all dress like bag ladies. It's the Harvard–Radcliffe look. I stopped in a local place for a bite to eat before I came up here. Not a single woman worth looking at." He stared out the window. With Charley, work was always combined with a search for some female companionship.

"Well, is that why you're here?"

"No. Met a lovely girl at the IWC. Some kind of deep-sea diver. Crazy about whales. Here's why I've come to visit. The Japanese want to harvest some of the biggest male sperm whales in the Pacific. They claim there's an excess number—more than are needed for successful reproduction. They claim that the best Western whale biologists say that not all the big males are necessary. Based on whale social behavior, there are many more males than are needed to mate with the females.

"A conservation organization approached me at the meeting. They said they would fund a study of sperm whale breeding behavior, to see whether the removal of some of the big males would make a difference. Here's the plan. We find out the details of whale social behavior. Then I will do an analytical model. You do a computer simulation. We create theoretical models of whale social and sexual behavior. Then we see if our results agree."

An analytical model is a pencil-and-paper mathematical model, the only kind that scientists and mathematicians could do before the invention of computers. By this time, I had had a lot of experience with computers, starting with the work in the radioactive forest, a pretty rare thing among ecologists at that time. When Charley and I were speaking, analytical models were considered the right thing to do in ecology, while computer simulations raised the eyebrows of both mathematicians and biologists.

Perfect, I thought, now we can do some real, hard science and help save whales at the same time. Just the kind of problem I had dreamed of back in the days I wandered around the radioactive forest and helped Heman Chase fix the old mill. Now I could make a difference.

Well, except there was one catch. Computer simulation—computer gaming—of something like the social behavior of whales seemed impossible. To most mathematicians, simulations seemed sham mathematics. To most biologists, computers were an unknown entity that seemed to have little to do with their work. I was one of the few ecologists at the time intrigued with the potential power of computers. Charley was sympathetic to my work, but he was a skilled applied mathematician and had approached problems of the conservation of marine mammals by clever, often simple but insightful analyses, using traditional pencil-and-paper math.

We talked that afternoon in my sparse office piled with books and papers, decorated only with a weary Christmas cactus, lonely in a corner against a pile of manuscripts on top of a series of drab brown file cabinets. My office was a long rectangle brightened only by sunlight streaming through windows on the northern side of the room, the side that faced "the Hole." Perhaps if we were to do something really useful with a computer about saving whales, then people in conservation would begin to appreciate what I was doing. Perhaps this would be a way for people in my field—ecologists and environmental scientists—to begin to appreciate the use of computers.

Still, it was a strange problem for me, trained in the study of forests, to tackle. I had studied with Murray Buell, an expert in plant ecology, who himself had been trained in plant taxonomy—the naming and classification of plants. But I had majored in physics as an undergraduate and I had always been interested in machines—electronic ones like the recording systems and computers at Brookhaven National Laboratory and the big mechanical ones at the old mill in New Hampshire. I had started to create computer models of forests and of endangered species.

"First, we have to find out what's known about the social behavior of sperm whales," Charley said. "We have to get together with those Australian whaling guys—the ones who think that doing science is rolling around in high seas in an open boat, banging into whales. Old what's-his-name—the one at the meeting last year—does a lot of field studies of sperm whales. Let's meet with him.

"We have to understand what in the social behavior affects the success of reproduction—mating, gestation, care of the young," he said. "You talk to who you can, I'll talk to some people in Britain. Make a computer model of the social behavior of the whales during the breeding season. Then look at the birth rate of the whale population as a function of the density of large males." Then Charley explained the financial arrangement with the conservation organization and other details of the project.

"Well, I'm off to see the lovely lady deep-sea diver," he said. "Hope you like the scenery here—all the cute houses and sailboats."

Then he left, and I sat in my office gazing once more out the window thinking what I would do next. I was elated by the idea that something like this might work, and I went for a walk along the shore. I always found it cleared my head to walk near the ocean. Good ideas would pop into my brain when I was watching the gulls circle or a boat approach a dock. But this day while I was walking, I saw Erene, who also worked with our group at MBL. She was one of the most beautiful women I had ever met, with ash-blond hair and blue eyes. She was recently divorced and so was I.

"You look very nice today," I said.

"I do? Thanks. You know, I try to dress neatly, but it's a problem here."

"How so?"

"Well, if I dress carefully, the other women think I'm a secretary. You're supposed to look sloppy here, but I just can't stand it."

It seemed a good idea to me that she dressed neatly.

After we walked along the shore, we wandered into town and ate lunch at the same restaurant that Charley had recently visited. It didn't matter to me what the other women in the place looked like; I was captivated by Erene. She told me she had taken the job at MBL as an adventure, but she really was a nester and had found herself a cozy apartment with a view of the coastal wetlands of Cape Cod, full of wildlife. It took me a while, circling around Woods Hole and walking with Erene, but before Charley came back to visit, Erene had begun to invite me and my two children, Nancy and Jonathan, to dinner.

Several months later, Charley came back and we met with a whale scientist, Richard Needles, another Canadian. "Here's how sperm whales' social behavior works," he said. "There's a pod of whales—the females and their young. During the reproductive period, a pod has a single male, the harem master. He mates with the females. Some pods haven't any adult males.

"Then there're adjacent males, ones on the breeding grounds, but without their own pod," Needles said. "They swim round looking for a pod without a harem master. Or sometimes they try to displace a harem master. Then there are other males—immature teenagers and others who just give up on mating—they go up to the Arctic feeding areas."

Charley and I visited awhile on our own and mapped out our strat' egy in more detail. "You should meet this diver lady," he said. "Has *no* fear. Goes down into the ocean in a kind of astronaut suit. Gone deeper than *anybody,* male or female. Crazy, but dresses great, looks great."

"Is this a study of your courting behavior or the whales?"

"Just like a family man," Charley shot back. "Stay-at-home. No distractions here. I know I can rely on you—nothing else to do here but work and watch the water and the clouds. None of your bloody big machines for me. I'm going to California. I can work on this on the plane and on the beach, then go diving, or snorkeling, maybe just swimming. You're the harem master, with two kids," he said. "I'm just an adjacent male. We have all the fun."

After Charley left, I sought out Bob Holmes at the Woods Hole Oceanographic Institution, one of the world's experts on whale sounds.

"How far can sperm whales hear each other when they call?" I asked.

"About fifty miles," he answered, but he looked at me quizzically. "Why do *you* want to know? I thought you studied elephants and trees."

"It's a long story—sometime over a beer," I said, and hurried away, anxious to avoid another embarrassing conversation about who and how whales should be studied.

With that, I thought, I had enough information to create a computer game about the courting and sex life of sperm whales. Here's how the game worked. Imagine a chessboard, only with many more squares—let's say a board about three feet by three feet, with hundreds of squares. The game has two kinds of pieces. One kind is a largish oval-shaped piece called a "pod" that covers about six squares; it's the group of females and their calves that Richard had told me about. The other is in the shape of a whale, small enough to fit in one square; these are the male whales.[3]

There is a small number of pod pieces—let's say twenty. You set up the game by throwing these on the board at random—say, by turning your back to the board and throwing the pieces over your shoulder onto it. Two players, one on each side of the board, have a pile of male whales.

Now the game begins. Each player takes one male at a time from his stack and puts it on the far right square in the first row. Then you, the player, throw three dice. The left die determines how many squares the whale may move up the board, the middle die determines how many it may move horizontally, and the right one determines if the horizontal movement is left or right—odd means go left, even means go right.

This way, a male whale piece moves completely at random until it gets within two squares of a pod. Then the odds change and there is a much greater chance of the whale moving toward the pod. The chances increase again when it gets one square away. That's because the male can "hear" the female and young whales in the pod.

You win a point by getting a male on the same square as a pod. Your whale can be displaced by one of your opponent's if his whale lands on a pod where you already have a male. There is some kind of dice rolling here, too, to determine whose whale gets to stay. Whales that end up on your opponent's first row have "gone to the Arctic" and are out of the game. The winner is the player who had the most males on pods for the longest time by the end of the game. You get one point for each play that a male is on a pod. A point is a new calf. So you win by having the greatest count of calves.

That's the general idea—the details differed in the computer game we created to more closely resemble the story about whale social behavior that Richard Needles had described to Charley and me. This would be a slow game to play by hand, but we wrote a computer program that created the board and the game and ran it very fast many times. We could play the computer version with various numbers of pods and males. Then we could make a graph of the average number of calves produced in relation to the number of males. When we created this game, we had a computer simulation of a little bit of the whale's social behavior as told to us by whale scientists. Based on information we obtained from Richard Needles, we added courting and mating of the whales to the game and the chances that all of this would result in the birth of baby whales.

We spent several months to get this program to work. As far as we knew, nobody else had ever tried to make a computer game about the courting and mating of big whales, or even about any animal's social and sexual behavior, not even human behavior. We were very proud of our work.

After a few more months, the computer game seemed to work fine. We began to get interesting results about how much the number of baby whales changed with changes in the number of mature males, both harem masters and adjacent males—the ones the Japanese wanted to hunt. The results were different from any we had seen written about before, and we thought we had the chance to write an important paper with a new insight about animal reproductive behavior. Just over the horizon lay an ocean of success.

But then a question arose and our poor computer model got lost in the waves. How long did a male have to be harem master—to be with a pod—for successful reproduction? Were these overnight stands—just long enough to mate successfully with each of the females? Or were these long-term relationships? Did the male have to be there for the entire breeding season to protect the pod? And what happened if one male came in and forced another out—divorce whale-style? I decided it was time to ask more questions of Richard Needles.

A short while later, Charley and Richard came to the Marine Biological Laboratory. Charley and I met together first to compare notes about where we were with our parts of the work. He had designed an elegant pencil-and-paper model. But because of the limits of that kind of math, he could only deal with fixed conditions about the whales and their environment. In the computer, we could have ocean conditions change during the game.

"How was California?" I asked as we walked from my lab office to the nearby cafeteria to meet Richard Needles. "Diving is scary as hell," Charley said. "But it was worth it to be with this lady scientist. The only thing is, she insisted I take a bath a day. We people of British descent don't like baths, you know—part of our social heritage."

We found Richard Needles seated at a table, drinking coffee. He was a large man, well over six feet, and built like a football player or a commercial fisherman. His face was tanned and starting to show lines typical of people who spend most of their lives outdoors, squinting into the sun and wind. He looked from Charley to me. It was clear that he didn't think much of what I was doing and he was skeptical about Charley, too, but knew him better from other meetings about whales.

"Saw the *Alvin,*" he said, referring to the two-person very deep ocean submersible research vessel. "Some good ships here. Ought to get you two out on 'em," he said. "Come to Canada. I'll show you what studying whales is really like. We'll go out in my boats. That's *real* science. Out in an open boat, watching the whales. Not the kind of garbage you guys do, sitting in an *office* scribbling on a pad, or even worse, using a newfangled computer. What's the world coming to, I'd like to know."

"We're making pretty good progress," I said. "We just have a few final questions." I cleared my throat and paused. "What happens if one harem master is chased out and replaced by another?" I asked. "Does the first harem master have to be there just long enough to mate with the females, or does he have to stay?"

"Oh," said Needles, "we can't tell one whale from another. We have no idea about that."

I was startled. "If you can't tell one whale from another, then how do you know any of the story you told us about their social behavior is true?" I asked.

There was a long silence. "Well," said Needles, "gazelles do it in Africa."

"That's it?" I said.

"Yes. We just assume that the whales behave like gazelles in Africa."

"You can't do that," I said. "We worked for almost a year to develop a computer model that mimics exactly what you told us. And now you're telling us you don't know if any of it is true. The whole thing's a god-damned fairy tale."

Needles got red in the face. I thought he might start punching us.

Charley looked away. "Science is more than rolling about in your bloody open boat," he said.

"Nonsense," said Needles. "We're out there where the action is. You guys sit inside and never see what you're studying. Now you're criticiz-ing me because you don't like what I tell you I saw."

"You just told us you didn't see anything—you made it up because you think whales are gazelles."

Needles got up from the table, mumbled something to the effect that people who use computers don't know anything about whales, and stormed away.

Our year of work on the computer model had just been hit by a tidal wave and sunk. The flotsam and jetsam of the programming was a beau no more basis in fact than the life of a unicorn. It gave fascinating results based on a totally mythical story told to us by a supposed expert on the behavior of whales. Except as an example of a hypothetical population, our work had no real application to the problem of hunting sperm whales.

I went back to my office, watered the poor thirsty Christmas cactus, and stared across the piles of paper that were, in part at least, the result of the development of our whale computer game. I looked out at the Hole, where sailboats came and went, and the ferry to Nantucket was arriving at the dock. I had a tremendous urge to get on it and leave those hypothetical whales roaming the computer ocean, forever lost and empty of meaning. And so ended my first—but not my last—venture into saving whales.

Several years later I told this story to my old friend and professional colleague Lee Talbot, one of the world's experts on conservation of nature. He had done some of the first fieldwork on big game animals in Africa years before. Lee laughed when I told him the story.

"I'm probably the culprit," Lee said. "I met with Richard Needles a while before he first talked with you, and I told him about the social behavior of gazelles I had studied in Africa. I told him about harems and harem masters, and how the male harem master defends his territory. He must have just transferred what I told him to believe it applied to whales."

It's hard to study whales. These huge animals are hard to find. Understanding has come a long way since that time. Ways have developed to recognize individual whales based on individual marks—cuts or markings on the tail, a certain shape to a jaw. Some whale biologists keep elaborate photographic notebooks, with entries about the sightings of individual whales. Perhaps today it would be possible to make a computer game that was accurate about how whales courted, kept their pods, and managed their young. But there are still the Richard Needleses of this kind of "science"—what I call the plausibility theory—if it sounds good it must be true. Informal observations substitute for careful measurements, the search for understanding, and the development of theories that are tested by observation.

I had become an active research scientist, still believing that I was part of a new but real science, a hard science, based on facts, solid scientific observation, and good theory. That theory was connected to observations, following the classic methods of science well-known in the modern world, in which I had been trained in my undergraduate major in physics. I still hoped that I could make a difference with that science, helping to apply it and really solve important environmental problems. But then I began to come across strange problems—like how many hours does a whale sleep—questions so simple that you would think only a child would ask, but questions I needed to know the answer to, and could not find.

So our wonderful computer game did not help save the whales and could not be used to increase our understanding of real whale behavior. In those days, nobody thought about software as something to sell; it was too arcane an activity to have much of a market, we thought. Perhaps today one of the computer game manufacturers could make an educational and entertaining whale mating game, a kind of animal pornography that might appeal to a certain audience.

I went for a walk along the shore with Erene, kicking my feet in the wet sand, listening to the herring gulls' raucous calls. We talked about sleeping whales. She laughed at my story and the world looked brighter. The ferry to Nantucket was leaving its dock. Tourists were everywhere. On the beach, kids dug in the sand. Young men strutted along the strand showing off their tans to the young women: adjacent males looking for mates, I thought. Charley Russell would say they were looking for a harem. The buildings of the laboratory—where so much research had been done for one hundred years, where famous scientists had studied animals and algae, where seawater could be pumped directly into the laboratories—in these buildings, scientists worked with microscopes and knew what they were doing. Perhaps, I thought, it was time for me to go elsewhere, to a place where there were forests that I was used to studying. It was time I got up my courage and asked Erene to marry me. She said yes.

A few weeks later, I got a postcard from Charley in Hawaii: "Having a wonderful time diving and seeing whales. Not too deep. Better than rolling around in some open boat." Perhaps his own courting had been a success.

Sometime in the future, I thought, this field of whale biology would grow up and people would be able to create useful theory well-connected to legitimate observations. Meanwhile, I would never forget the story of the social behavior of sperm whales, based on what gazelles did in Africa.

Perhaps I really had gone into the wrong field. Where was the science in "ecological science"? It was lost at sea, rolling about in an open boat.

Eight

🙠

CIGARETTES AND THE
SUMMIT OF MOUNT WASHINGTON

e were hiking in the White Mountains of New Hampshire, in its Presidential Range, that have the tallest peaks
in the northeastern United States. We were above timberline when the rain came so hard that it ran down our
legs in sheets, filling our boots with water. It no longer mattered if we
walked in a stream or alongside it, so I chose to walk in the stream where
the going was easier; the rocks and pebbles were formed into a smoother
line there by the moving water than the jumble of rocks alongside.

We were in a saddle between Mount Washington and Mount Madison. I had stood at the same point on a clear day and could see sixty-five
miles to the patches of woods and farms below. In front of me was a
wonderful view of the summit of Mount Washington, the tallest of the
Presidential peaks. On clear days, the tundra above timberline had a special charm: tiny tufted plants huddled in spaces between loose boulders
that had been cast haphazardly about ten thousand years ago by the

great glaciers. Now the boulders were reduced to lichen-covered rocks that rang when you stepped from one to the other, almost musically. White-throated sparrows called and fluttered from tundra tuft to tuft.

But on this day we could barely see to the next cairn marking the Appalachian Mountain Club trail. It was midafternoon on the third day of our hike. Eight of us had started out on a lazy, humid summer afternoon at the foot of Mount Madison, some of us from Yale University and others from the University of Vermont. It was meant to be a botanizing trip to explore the high elevations above timberline, enjoy the scenery, and try to identify and learn the plants of the mountains. But as was common on these mountains, the weather had turned bad.

Mount Washington gets some of the worst weather in the world. It is a curious mixture of the wild and, at the summit, the humanized. Near the summit there is a permanent weather station manned year-round, the terminus of a cogwheel railway that takes tourists from the base to the top, a small coffee shop connected with the railway, and, not too far away, one of the Appalachian Mountain Club huts where hikers can stay out of the weather and obtain meals. A young friend of Heman Chase's had hiked up Mount Washington on one of its wilder trails. When he returned and visited the old mill, Heman asked him what the hike was like.

"Pretty good," the young man said. "But you know, after we climbed all day, made our way up the headwall of Tuckerman's ravine, sometimes hand over hand holding onto roots and branches, then walking in a stiff wind and thinking we were really outdoors, we came upon this railroad station where there were some old ladies waiting for a train. Kinda took the adventure out of it."

Mount Washington is famous as the location where the highest wind speed ever recorded on the ground was observed. It once blew 232 miles an hour. Three major weather systems that travel across North America meet at Mount Washington: one that swings down from the Canadian

Arctic over the Midwest and the eastern United States; another that comes across from Washington–Oregon–Northern California; and a third that brings moisture up from the Gulf of Mexico. It has snowed on Mount Washington in every month, and the weather can turn quickly from a warm, T-shirt sunny day to hail and freezing rain with temperatures near freezing. Knowing this, we were prepared, but it wasn't a day to be strolling along, examining tiny flowers through hand lenses or gazing at distant vistas.

We reached a decision point. The trail divided. One trail went up to the summit of Mount Washington, the other followed the middle of the saddle down to the road where our cars were parked. The eight of us stopped and talked about what to do, a discussion in the pouring rain. I had the feeling that if somebody were watching us from afar, they would think we were a little out of our minds to be standing in this pelting rain having a calm discussion about what to do next, the water pouring in sheets down our rain gear. And suddenly plain human nature came into play, bad weather or not. Competition among people and the challenges between the sexes started to interfere in our abilities to get things straight and take the wisest path. This hike had begun as a trip of professional botanists and ecologists who had agreed to take a relaxed pleasure trip through beautiful country. It was an extension of a practice that Jim Jacaranda and I had dreamed up at the time. We were on the faculty at Yale University. Tired of being stuck in university offices and laboratories all day during much of the school year, we organized "forays into the environment," the silliest name we could think of for an excuse to get out and walk and look at the landscape and its life. Our purpose was to learn about the plants and their habitats and enjoy ourselves.

But in the saddle below Mount Washington, in the rain, one of our eight was a young woman on the faculty at the University of Vermont who had been outdistancing everybody the entire trip, walking faster and longer and identifying plants quicker than anybody else. Some of the men took this as a macho challenge; she was not going to outdo them. I don't think anybody except this young woman was especially keen to make the ascent to the summit of Mount Washington, but there was a si-lence among the party as the rain poured down, the wind blew, and rolling clouds obscured the mountain peaks. The trail was well marked by cairns—each a small pile of stones—set about a hundred feet apart, built over the past century by volunteers of the Appalachian Mountain Club. The rain was so heavy and the fog was so dense that we could just see one cairn ahead. It was wild weather in a wild landscape, however much people long ago had built beautiful paths within it.

My decision was simple. I had come to enjoy myself and learn about the alpine plants. This was not a day to do either. On many previous trips, I'd been to the summit of Mount Washington. On this day, it wasn't fun and I had nothing to prove to myself. I was going to take the path down.

But the other men waited to hear what the Vermont woman was go-ing to do. There was a long and ominous silence. "I'm going up the moun-tain," she said, and started out into the fog on the trail that rose upward. The other men hesitated. If things got really bad, they rationalized to each other, they could take the cogwheel railway down from the summit.

Three of us, myself and two women, agreed that we would take the descending trail, get some of the cars, and meet the others at the base of the railway where there was a parking area. Then we would drive the wet hikers back to their cars.

We parted and, as I waded down the streambed with the water soak-ing through my raingear, I thought about the five who had continued up Mount Washington. I hoped they would be okay, imagining many kinds of disasters they might face. A gust of wind might blow one off a cliff. Soaked through, some of them might suffer hypothermia.

Our route down the mountain was hard enough. In spite of our rain gear, we were completely soaked through when we descended to the elevation where trees grew—about four thousand feet above sea level and a thousand feet below our parting with the rest of the group. I had hoped the trees would make travel a little drier, but I kept brushing up against limbs and twigs that spilled more water onto me than the rain had flung at me higher up. White-throated sparrows continued their calls, a cheery sound cutting through the fog. In its own way, travel through the storm had its own fascination—a different kind of contact with nature but, with twilight approaching, one that we were happy to see end after several hours.

Eventually, wet, tired, and hungry, we three reached the base of the mountain where our cars were parked. We changed into drier clothes and drove around the mountain to the base of the trail that our friends were to take down the mountain.

Near dark, our friends arrived. Jim Jacaranda, one of the five, and the most open and honest about his own failings and limitations, told me what had happened on the way up to the summit and back down. They reached the summit house of Mount Washington in midafternoon and went inside for a cup of coffee and snack and to make a decision about the descent. They were warmed by the little patch of civilization on the top of one of the wildest mountains in America.

The engineer of the cogwheel railway sat nearby, drinking a steaming and inviting cup of coffee. They asked him if he had room and how much it would cost to take the cogwheel train down the mountain. He said he could take them, but they had to make up their minds quickly. "Last trip of the day," he said. They looked outside. The weather was getting worse, with the rain coming in sheets at a sharp angle to the ground, blown by high winds. Light was fading as the afternoon drew on.

None of the men dared say a word. None would be outdone by the woman from Vermont. Jim and the other men looked forlornly out the window. Bill, one of the younger men, stood up and spoke. At first he said he had decided to take the train but hesitated when nobody else offered to join him. In the silence, he changed his mind. He would not be the only one to give in.

The Vermont woman said *she* was hiking down, whatever the others did. The rest watched the engineer go outside and climb into the cab of his train. They watched as the train disappeared down the mountain. Jim said the three other men looked forlorn; the woman looked resolute.

"The hike down was terrible," Jim said. "We took the Abel trail. It was very steep, and water poured down the slopes like a waterfall. The wet ground was slippery and there were places it would have been easy to fall. We passed a plaque that we stopped to read, clearing our glasses so we could see in the dim light. It was in honor of somebody who had died while climbing that trail—a little creepy, given the weather," he said.

Finally, close to dusk, the party made it out and walked over to the lower train station. The Vermont woman led the way and the men struggled to keep up with her. None could.

When she reached the parking lot, she went past the parked cars in the rain to a kiosk, still going strong. The men followed, still not willing to let her get the best of them. She went over to a cigarette machine, opened her backpack, took out some money, and bought a pack of cigarettes. Then she opened the pack, struck a light, and began to smoke.

"I thought all the other men were going to buy a pack of cigarettes as well and start smoking, just to prove they were as tough as she was." Jim laughed. "And none of them were smokers, either," he said with a chuckle.

What had started out as a pleasant, leisurely hike had, for them, become a gender challenge. They were wet, tired, and miserable, and they hadn't enjoyed themselves, but they had the pitiful satisfaction that they were not completely outdone by the Vermont woman. She kept ahead of them, but they had done everything she had done until they reached the cigarette machine. There, she finally outdid them. Even if they were smokers, they would probably have been too exhausted to light up. And if they smoked one cigarette, she would probably light up a second. She had beaten their spirit; she had won. It seemed sad to me that the macho attitude—or any hang-up that forces one to do what he does not want to do—had dominated the enjoyment of nature. It happened then and it happens now. Even in this kind of physical activity, some can't let go of competition when it surfaces. So, scientists who try to study the great outdoors are just people like the rest of us. Being scientists doesn't raise us above any typical human failings. But this kind of intense, emotional competition can get in the way of finding out what is actually happening to nature and to ourselves.

I go to the woods to enjoy myself, to feel better, to be uplifted in spirit. When the weather turns and it looks like it's time to let discretion rule over valor, I remember the cigarette machine at the base of Mount Washington and the young woman outhiking and outsmoking some tired men. I remind myself about the reasons I go to the woods: to uplift my spirit and to learn a little more about nature. If I want to suffer, I can do that easily at home. If I want human competition, I can find that easily in the valleys, in the cities, among competitive professionals, within government agencies. I imagined the same hike if I had taken it with my father-in-law Heman Chase, or his gentlemanly brother-in-law, the quiet New Hampshire alcoholic, Walter Burroughs, or with Clarence Goodenow. I imagined Elsie Goodenow asking him what it was like outside and Clarence answering something like "Could've been better." But then, none of them would have made the trip in this weather. It would seem to be unusually poor common sense to do so, something only those strange scientific fellows might do or young men out trying to find themselves or prove themselves. Lee Talbot, the gazelle expert, might have described it as adjacent males trying to be harem masters.

Nine

WEIRD

 was standing on a pleasant path between two rows of trees, on the grounds of the IBM Thomas J. Watson Research Laboratory in Yorktown Heights, New York, dressed in my ecologist's field clothes—blue Sears and Roebuck workshirt with many buttoned pockets; khaki trousers; heavy leather hiking boots with Vibram soles; a compass hanging round my neck on a long leather shoelace; and a large aluminum-frame backpack held on by wide straps around my shoulders and waist. I spoke about the trees and about ecology, looking at a camera set up on the other side of one line of trees, along with a cameraman, his helper, a soundman and his helper, the director and a producer, and a few extraneous assistants and onlookers. There was enough equipment to install a telephone line, highly visible from the path.

It was strange work for a field ecologist, but IBM was making a television advertisement about my research, done in cooperation with two excellent IBM scientists, James Wallis and James Janak. We had created a computer program that grew trees in a forest and for which I had searched to learn the number of leaves on a tree. This research had be-gun as an IBM-sponsored summer study to show that computers could be applied to socially relevant issues—in my case, environmental issues.

Most of the four days of filming had taken place at the Hubbard Brook Experimental Forest in the Presidential Range of New Hampshire—the same mountain range where the botanists and ecologists chased after the cigarette-smoking woman professor. Hubbard Brook Experimental Forest was then and still is one of the best examples of research in ecol-ogy in the world. It began as an idea of two ecologists, Herb Borman and Gene Likens, then both at Dartmouth College, and Robert Pierce, a U.S. Forest Service scientist in New Hampshire. The first major experiment at Hubbard Brook was clear-cutting an entire watershed—about a hun-dred acres—and comparing the effects on water quality and total runoff with an uncut watershed nearby.

IBM wanted the advertisement to be completely accurate in content and in personnel; anybody who appeared in the film had to actually work on the project and the content had to completely and accurately re-flect the work. During the four days we spent in the forests of Hubbard Brook, a professional film crew filmed everything—rain showers, changes in the lighting of the forest as the day passed, every piece of field equip-ment in use by any of the researchers, a person's hand moving over a topographic map of the forest. The director told me that it was much cheaper to film everything anyone could think of than to discover after-ward that something crucial had been missed, which would entail trans-porting the entire crew back to New Hampshire. They shot 7,000 feet of film for an advertisement that used 270 feet, he told me, and this was a long ad; it ran, in its initial version, for almost three minutes.

In spite of all his care, the director decided later that he had not filmed everything. We had one day more of shooting, which was to be done at the IBM Thomas J. Watson Research Center in Westchester

County, north of New York City. We were supposed to spend the day filming our work with computers. But a few days before, they called and asked if I could bring my hiking clothes and ecology field equipment so they could film some more material with me talking about the forests.

About three in the afternoon on the day we had been filming inside the research facility, I changed into my field clothes—as I mentioned, hiking boots, broad-rimmed field hat, blue long-sleeved, many-pocketed field shirt, and tan trousers. I put my large backpack on and walked out into the large and pleasantly planted landscaping around the research center. The center had been built on an old farm, and there were many acres of open land. This made a pleasant environment in which people worked on all kinds of exotic science that involved some kind of computing. A long grassy path led from the research center to the main road. The path was lined on each side with trees. The film director decided that if he put the film crew outside one line of trees and if I stood on the grassy path with the other line of trees behind me, the scene would appear as if it were in a forest. In this way, he could avoid paying for the entire film crew and myself going back to New Hampshire.

The advertising company wanted to compensate me for all the time I had taken to participate in the filming. However, this was a little complicated to do, since the film was about my work; therefore, they could not pay me directly as an expert or consultant because that would compromise my independence as a scientist. However, they said that they could pay me as a professional actor, and they did pay me at the going rate of the Screen Actors Guild. I felt strange enough to be in my field garb on the lawns of one of the highest of hi-tech research centers. Since I had absolutely no ability as a thespian, this stint in which I was paid as a professional actor made the situation all the more peculiar. It was a combination of Heman Chase's style of life, the nineteenth-century village and the late twentieth century—like Brookhaven National Laboratory—but thrown together in a way I had never expected. Once again the images of Heman, Walter Burroughs, and Clarence Goodenow came to mind. But unlike Walter, I hardly felt all right for my age and habits. I thought that Heman would have frowned at the willingness with which I engaged in commercialization of my work, not only lending my name to an advertisement but accepting pay as a professional actor. Or would he have frowned? He often wrote and spoke about what he did, and I read what he wrote and listened to his stories. Generally, he argued in favor of the simple life, but he had a flair for storytelling and loved the theater. Well, maybe he wouldn't have minded. I didn't know, but I was thinking about it. I tried to push these thoughts from my mind and focus on my lines, a very difficult job for me.

The filming required quite a number of takes, in part due to my non-existent acting abilities. In the middle of all this filming, a man walked out the door of the research center and headed down the path toward us. We were sure he would see the film crew and stop or go in some other direction, so we just kept filming and I kept talking. But he did not stop. In fact, he walked quickly through the scene while I continued to talk and the camera kept rolling. He looked straight ahead and walked right past all of us, seeming not to notice anything that was going on. That was one take we had to do over. We were completely baffled that the man, who had to have seen the many people in the film crew and their elaborate equipment, was impolite and did not stop until we had finished.

This was so weird that I mentioned it a few days later to one of the IBM scientists I was cooperating with, Janet Lyne. A few days afterward, she called me to tell me a curious thing had happened. She said she had just eaten lunch with another member of the research center's mathematics department and he had told her about a strange person he had seen on the IBM grounds a few days before. He said there was this man wearing a backpack and hiking clothes standing on the footpath right near the building, looking at the trees, and talking very loudly to himself. "I walked by him as quickly as I could, I'll tell you," he said to Janet.

He had seen me standing in the path from afar, and the sight of some-one dressed as I was and talking loudly led him to assume things about me that made him miss seeing the film crew and all of its equipment, even though they were completely visible from the path and their equipment was not quiet.

I was reminded of the IBM filming experience years later when I made a trip to British Columbia and was taken out to see some modern forestry practices. The night before, a number of Canadian foresters met me for dinner. The dinner conversation turned to a discussion of odd be-havior. One of the Canadian foresters said that the strangest story he knew along those lines happened to a friend of his who was a rancher in the dry eastern part of British Columbia. This friend would make occa-sional business trips to Vancouver. Once, he flew to Vancouver and for-got to take any rain gear. He arrived downtown with only his cowboy clothes—a big Stetson hat, blue jeans, boots, the whole works. He was walking after dark in what he considered a bad part of the city when it began to rain hard. All he could do was pull his Stetson down tight on his head and keep walking. Up ahead he saw a lighted corner where, un-der a store awning, stood a group of punkers with Mohawk haircuts, purple-dyed hair, tattoos, cut-off T-shirts—typical dress, he thought, of some of the tougher and more dangerous city gangs. So he was immedi-ately wary.

Used to dealing with dangerous animals on his ranch and when he went out hunting, the rancher knew how to approach potentially threat-ening wildlife: walk calmly, try to show no fear, and look no predator in the eye. So he took a deep breath and proceeded as slowly and calmly as he could, head down, not staring anyone in the face, but keeping a care-ful watch out of the corners of his eyes. As he passed the punkers, he said, "Nice evening"—part of his attempt to demonstrate that he was not afraid, although he was. Nobody answered him. But as he was passing the group, one of the punkers looked at him and said to another, "Weird, eh?"

Each side had its preconceived notions about the other. The rancher assumed the punkers were dangerous, and the punkers believed that anybody walking down a street in Vancouver dressed as a cowboy, hat pulled tight over his head, and without an umbrella or raincoat, was at least a little crazy. The IBM mathematician thought I was crazy while the film crew and I believed the same of him. So it is with many of our perceptions, even affecting how we perceive and deal with our environment, with nature, with all of life. Affected in this way, we continue to confuse ourselves about who is normal, who is not, even who is an expert. We fail to do right for our age and habits. And sometimes we have to admit that we do act quite strangely.

Ten

~

THE SEARCH FOR THE AMAZING
TRIPLE-CANOPY RAIN FOREST

 was searching for the famous triple-canopy rain forest—
the kind that I had seen depicted in popular magazines,
discussed on television nature programs, and described
in scientific literature. Rain forests were supposed to
have the greatest diversity of life of any kind of landscape. They also
were supposed to have a special and wonderful vertical layering of three
kinds of trees: trees at the top that shaded all below; an understory of
smaller trees struggling to reach the bright sun shaded from them or able
to persist within this secondary position; and a third lower layer of even
more light-suppressed and shade-tolerant plants.

Four of us struggled down a muddy slope in the cloud forests of
Monteverde, Costa Rica, one of that country's most famous rain forests.
A light drizzle wet the ground and all surfaces, making everything slippery. With me were Lloyd Simpson, who helped the lady with the broken arm at Paradisio Perdu; Mike Marzolla, who was also with us on that

trip, and my wife, Erene, who, since we met and married at Woods Hole, often traveled with me. She too wanted to see the forests of Costa Rica.

The path was narrow and the temptation was to grab onto the trunk of one of the many small trees to prevent oneself from falling down. But this could be dangerous. Some of the trees were armed with thorns over an inch long, ending in a point as sharp as a needle. The thorns extended from top to bottom of the tree. A grab with one hand would bring you in contact with a dozen or more. Also, tree trunks were favorite hiding places for some poisonous snakes. The bark of the trees contained some of the highways for bullet ants, ants about an inch long, so-called because their bite was a sting that made you feel that you had just been shot with a bullet. Paulina Petrova, my friend and colleague, was taking us to see her research sites on the wet, misty slopes high in the mountains. She was one of the world's experts on these forests and moved quickly and smoothly down the trail.

"Don't worry about falling," she encouraged us. "The way you become a tropical rain forest expert is to fall down a lot." Soon we believed we were among the world's greatest tropical rain forest experts.

The struggle down the trail was the realization of one of my dreams, to see this famous, high-altitude rain forest. Rain forests have a special mystique among naturalists, botanists, and ecologists. They are famous not only for their great diversity of species, but also for that complex architecture known as the "triple canopy," which, as I mentioned before, was three vertical layers of trees extending far above the ground. For years, I had studied forests in North America and had long wished to see the tropical forests of Central and South America. That chance came when I started a new research project on tropical rain forests and was searching for the best location to do the fieldwork. As part of my search, I traveled to Costa Rica and then to Manaus, Brazil.

Lloyd Simpson, a professional forester working with me, was going to direct the fieldwork we would do and wanted to see how difficult that work would be. A football linebacker in high school and college,

Lloyd was big and strong and had hiked through hundreds of miles of remote country as part of our research, traveling through the far north of
Canada, Alaska, and Siberia. Together, the two of us were familiar with
forests of many kinds, except for the tropics.

Our hike with Paulina was our first view of tropical rain forests in
Costa Rica. She was studying epiphytes, plants that cling to trees, and
she was famous for her method of reaching these plants. She used mountain climbing techniques to ascend tree trunks. Up in the trees, she found
a unique, miniature world among the epiphytes and the high branches.
Entire tiny communities of plants and animals lived high up in the canopy,
existing on the rain from above and on the nutrients that were leached
by the rain as it passed through the leaves. A trim, athletic woman with
short dark hair, Paulina moved gracefully and quickly through the narrow, muddy trails, speaking enthusiastically about her work and the
wonders of the tropical forests as she disappeared in the mist and drizzle ahead of me on the trail.

Monteverde, in the mountains of central Costa Rica, was settled in
the 1950s by Quakers from Alabama who cleared the slopes for dairy
farming, providing today one of the major sources of milk products for
the country. The Quaker settlers intermarried with Costa Ricans, and
their English- and Spanish-speaking children grew up during the begin-
nings of worldwide environmental awareness of the late 1960s. Part of
the farmland was set aside as a nature preserve; it is now a major tourist
attraction, famous especially for views of the quetzal, a rare mountain
bird with a long and exquisite tail. Scientists were attracted to the preserve, and Monteverde is now a center for research and education about
tropical rain forests. What could be a better place to begin to learn about
these forests, I thought, and especially to see for the first time the famous
triple canopy, which I had heard about for twenty years?

The triple canopy is said to be one of the main distinguishing features
of this kind of forest. Textbooks show diagrams of rain forests with
three distinct layers. Photographs taken from above by people in bal-

loons and helicopters show the top layer as an almost uniform, continuous, dense sea of green with occasional trees emerging like islands above
the rest. In general, the tips of neighboring trees do not quite touch. The
triple canopy is supposed to be a marvelous, beautiful display of nature's
bounty. Green, green, and green—layer after layer.

Tropical forests also are said by biologists to be ancient. As a result of
this belief, tropical rain forests are supposed to epitomize another persistent belief—that forests, if left to themselves without human interference, would achieve a maximum amount of organic matter and number
of species. According to this belief, the lack of human interference, the
ancientness of the trees, and the triple canopy made tropical rain forests
extraordinarily rich in species. Popular and scientific articles about these
forests gave me the impression that I would find a magical kingdom of
ancient trees in comparison to the meager forests of North America that
I had studied for thirty years.

As I stumbled along the narrow trail, trying to keep up with Paulina
but also stopping frequently to look at the leaves and twigs and wet tree
trunks, I was surprised to discover that the forest was not at all like the
pictures and diagrams in books, or the accounts in scientific articles I had
read. The forest was quite open, with big breaks where trees had fallen
down. The trees were not especially big, and there was only one layer,
not the triple canopy. And the forest lacked the mysterious darkness said
to be characteristic of the tropics.

When we stopped to look at Paulina's research area, I mentioned
these thoughts to her. She said that these high-elevation forests of Monteverde were not typical of tropical rain forests. We would have to visit
the lowland forests to find the triple canopy.

So we continued our search for the triple-canopy rain forest. On our
way back to San José, Costa Rica's capital, we stopped at Carara, a small
nature preserve that contained the kind of lowland forest Paulina said
would have the triple canopy. It was near the port city of Punta Arenas

and therefore near to sea level. Although it was hot and steamy, we hurried into the forest, hoping to get our first view of the famous ancient
and triple-canopy rain forest. But we were disappointed again. We were
surprised to discover that Carara looked a lot like what we had seen at
Monteverde. There were many breaks in the forest where trees had
fallen. Sunlight shone through brightly, and the gaps were dense with
the green of many small young trees and vines. We examined broken
stumps of fallen trees and saw that the wood was very soft; the trees
would blow down readily in a storm. These were young trees, not ancient ones. And there was, once again, just one layer of trees.

Coming out of the forest at Carara, we met a group of students and
their professor, who had taken them to see the forest as part of a course
in forest ecology. He was a professor of silviculture from the University
of Costa Rica and was one of the nation's experts on rain forests. He
agreed that Carara did not have a triple canopy and told us to go to the
eastern side of the country, near the Caribbean coast, where there were
much more extensive rain forests than at Carara. Along that coast, he
said, we might find the triple canopy.

A few days later, we traveled to a research station called La Selva,
which means "the jungle," located on the eastern side of the country, not
far from the coast. La Selva is a research and educational center operated
by the Office for Tropical Studies, an organization run by a conglomerate of American universities. There we found huge trees with buttresses,
a thickening of the stems from the ground up to ten or twenty feet. Although the trees were larger than we had seen at Monteverde or Carara,
the forest had many breaks and openings, many fallen trees, and it too
lacked the triple canopy. We asked the botanists there about our quest.
They told us we would have to go to the Amazon Basin to see the triple-
canopy rain forest—that it was not really characteristic of the forests in
Costa Rica.

So on we went. A few weeks later, Lloyd and I traveled to Manaus,

Brazil, the famous city on the Rio Negro in the Amazon Basin, just above where this river joins the Solimões to form the Amazon River, nine hundred miles upstream from the ocean—the major jumping-off place for those who wanted to visit the interior forests of the Amazon Basin. We were taken into Amazon forests by forestry experts from the Brazilian Amazon Research Agency, known locally as "INPA," the acronym of its Portuguese name. Both American and Brazilian nationals worked there. They were said to be the most knowledgeable about these forests of any people in the world.

Christopher Ortega, said to be the most important authority on the Amazon forests, took us to an experimental area where he was studying the effects of clearings made by farmers. A field crew was felling trees, weighing them, and measuring their diameters and heights.

The forest was made up of small trees in fairly uniform stands—again, no triple canopy. The forest reminded me more of woodlands in Virginia and Connecticut than the storybook diagrams I had seen of tropical rain forests. The trees were no bigger than you would see in New England or the Atlantic coastal states of the United States. Once again, within the only one layer of trees, there were many breaks and openings. There were many, many species, but to a casual visitor like myself, most of these trees looked similar to one another because they differed primarily in their flowers, not in the shapes of their leaves or color of their bark. Christopher told us that the kind of forest we were looking at occupied 90 percent of the land area in the Amazon Basin. The photographs of huge Amazon trees, he said, were taken in the forests right along the river. Those riverside forests occupied about 5 percent of the Amazon Basin. Even in Brazil's Amazon Basin, we had failed to find the triple-canopy rain forest.

The next year, having decided to do our study in Costa Rica, we began a cooperative project with a number of organizations in that country. One cooperator was Margules Profundo, said by other scientists in

Costa Rica to be one of the field scientists most knowledgeable about rain forests of that country. Margules worked for the Lentico Corporation, a company that made doors of tropical hardwoods and was attempting to incorporate sustainable forestry practices on its lands. Margules took Lloyd and me to a rain forest on the Caribbean side in the lowlands of Costa Rica, where the professor of silviculture we had met at Carara told us the triple canopy might exist.

It was the rainy season but the day was sunny and so hot that the warm air steamed with moisture. We walked through the steaming woodlands, stepping carefully around large pools of standing water and over fallen logs. Again Lloyd and I looked in vain for the elusive triple canopy. Instead, we found a patchy, open forest with a single layer of trees and with big breaks where trees had fallen over.

After we explored the forest, we came back out to the dirt road and our car. Hot and sweaty, we tried to cool off by sitting under the shade of several balsa trees that reached thirty feet over our heads. We ate our pack lunches and talked with Margules. I said we'd been searching for the triple-canopy rain forest that we'd read about, but had not yet found it. Margules said, "You have, too? I've been looking for it myself. I've looked all over the world. I couldn't find it here, and I took a trip to Malaysia and I couldn't find it there, either.

"The trees in these forests grow fast," Margules said. "Their wood is very porous and weak, and storms easily blow the trees over. Look at the balsa wood trees we are sitting under. When I came here six years ago, these trees had just sprouted. Now they're thirty feet high. That's five feet of growth a year." (Balsa wood is so light that it is used in model airplanes.)

"Something else," Margules said. "Most of the trees in these tropical forests are not tolerant of deep shade. They grow poorly if at all in the deep shade of a really dense forest. Their seeds germinate only in open conditions. In these forests, trees grow rapidly so they can gain a place to capture sunlight before other trees beat them in this competition. Such fast-growing trees aren't likely to produce a long-lived canopy but instead a canopy that breaks up quite rapidly." I mentioned that Christopher Ortega had told us that most Amazon forest trees had an average age of forty to sixty years.

On returning to the United States, I picked up a recent copy of *National Geographic* magazine. There was an article showing Paulina Petrova's husband, Jack, standing on a limb high up in the rain forest of Monteverde, his head peering out over the canopy. The article, written by a well-known biologist and not by Paulina or her husband, described rain forests as having an upper canopy with emergent trees, a lower layer of trees "specialized for this twilight region," and "still lower" a layer of saplings and herbaceous vegetation. The article claimed that the photograph showed the triple-canopy rain forest.

Next, I checked a recent technical book, *The Tropical Rain Forest: A First Encounter,* to see if I missed anything in my previous readings. I found a confused message. At the beginning of the book, the author, Jacobs, wrote that it is often possible to "discern several layers" in the canopy, but near the end of the book he denied its existence on the basis of one recent study. In other words, he equivocated.

And so the myth of the triple-canopy rain forest continues. Like El Dorado of the Spanish explorers, the triple-canopy rain forest seems to be a search for a mythological kingdom. People I speak to continue to suggest new places to look. I get the feeling that somebody is always willing to point me in a new direction as long as that direction is away from where they happen to be. Some say that it exists in the West African rain forests of the Congo River Basin, where nobody I have spoken to has seen it. They say, however, that these forests led to the drawings that are in all the textbooks.

So it is with so many beliefs about nature. Preconceived notions often dominate—to live on, even when contradicted by the facts. The search for the amazing triple-canopy rain forest is just one illustration of

how we tend to deal with the environment through ideologies, mythologies, folk wisdom, and folktales, as much or more than we deal with it by
direct observation and fact. We live with many myths about nature;
often these are told to us as if they were true. We hear about them on
television nature programs and in popular articles in national magazines.
Sometimes it is difficult to tell the difference between one of these
myths and the conclusions of scientific research. Perhaps the most surprising aspect is that a myth sometimes is perpetuated by scientists in
spite of growing evidence to the contrary. Even scientists filter what
they believe they see through a blinder of cultural mores and mythologies.

The story of the triple-canopy rain forest composed of ancient trees
would be a fine tale if that's all it was. But our policies, laws, and actions
for conservation are determined by these stories. The real world im-
poses itself on our false attempts at management. While the facts seem
to be quite different from the myth, we seem determined to believe in
ideas like the triple-canopy rain forest—in spite of facts, not because of
them. Facts, like the real age of Amazon Basin forests and the real structure of these forests, interfere with such myths. In my thirty-five years
of experience studying ecology and trying to help solve environmental
problems, I have found that policies developed on myths and folktales
have led to one failure after another for natural resources. I suspect the
same will be true for the tropics.

We may like to search for the El Dorado of the triple canopy, but if
we really want to conserve nature and understand it, as well as understand ourselves, we have to get beyond our pleasing preconceptions and
look at the rain forest directly, drizzle or not, as one slides and slips
downslope. We must take the tropical rain forest for what it is, just as
we must understand all of nature for what it is. We must look, observe,
record, count, and study nature without the blinders of ideology, myth,
and folktale. And perhaps one day, one of my colleagues more expert in
tropical rain forests than I will finally discover the reality about the
amazing triple-canopy rain forest.

Eleven

HOW MUCH FOOD
DOES AN ELEPHANT EAT?

oger Atwood climbed up on the wing of a Cessna 182, a high-wing, single-engine airplane, opened a five-gallon fuel can with a bottle opener, and poured gasoline into the wing tank. We had landed on a grass strip in the middle of Tsavo National Park, at the time an essentially abandoned landing strip in the midst of a largely unvisited and unused park covering 5,000 square miles—larger than the state of Delaware. A hot wind blew sand and dust into our eyes. I stood on the ground, handing cans up to Roger. We had spent three hours flying over this huge park and had another three hours to go. We were stopping to refuel and to eat lunch.

Roger had volunteered to fly me over Tsavo to see whether any wildlife still roamed its warm and dry landscape and to see the effects of a major drought and a resulting huge elephant die-off that had taken place a decade before. An estimated six thousand elephants died during that drought. They died of starvation, not from lack of water. The vegetation gave out in the drought, and the elephants and other wildlife ate most of the low-lying vegetation. The elephants then debarked and knocked over many of the trees, feeding on bark and leaves, finally succumbing to starvation. Next to human beings, elephants are the greatest land clearers of all the mammals.

I had come once again to Africa, this time to begin a research project about elephant populations and their ecosystems. I had come to the study of elephants by a circuitous route. Once people and organizations had involved me in helping with endangered species, I began to think about how one might act in advance—to predict when a species might become endangered before it actually did so. What we needed was a species that had the characteristics of those that tended to become endangered easily, but was not yet so, and one for which there was a lot of information. We could study this species and perhaps develop some kind of forecasting tool.

Species that are long-lived and have relatively few offspring over time tend to become endangered more easily than species like rats and cockroaches that live a short time and have many offspring. The African elephant seemed to be the perfect animal for us to study. At the time, they were still abundant but were threatened locally, and unlike so many other cases that had been brought to me, elephants had been the subjects of some good population estimates. So we began a study of elephants in Africa. First, we drew up some straightforward questions about the potential rate of increase of elephants. But we quickly realized that elephants were strongly affected by the availability of water to drink and vegetation to eat. So we now were involved in a study of how changes in the supply of water and food might affect the chances of elephants surviving while they were also under pressure from poachers who hunted them for their ivory. We obtained some research grants to do this work.

Roger got back into the pilot's seat and we taxied out onto the grass strip and took off to the south, banking westward in a large circle that would take us over the western border of the park. As we flew, Roger told me about his experiences over the years with elephants. A long debate had ensued in East Africa about how best to conserve and manage elephant populations. Poachers were killing them and stripping them of their tusks to sell them on an international market, leaving the rest of the carcass to rot. This threatened to reduce the elephant populations to a dangerously low level. But protection of elephants in certain areas had been so successful that the animals were locally too abundant and were eating out their food supply. The most famous case was Tsavo's elephant herd. A rapid buildup of elephants after Tsavo became a park was in large part the result of the activities of David Sheldrick, its first head, who sought to stop poachings and provide year-round water for the elephants by digging artesian wells and damming the park's largest rivers. These efforts resulted in a population increase so rapid that there was much concern about a possible die-off during the first drought.

A debate ensued. One side said that the only solution was to cull out some of the animals—shoot them to reduce the population to a manageable size. The other side argued that nature should be allowed to take its course. Nature would achieve a balance that would satisfy everyone.

Roger told me that he had tried to help control elephant populations where they had become overabundant. Poachers killed elephants in the most depraved way—taking only the largest animals with the largest tusks and not utilizing anything else. Roger and others believed that this approach not only was cruel for the individual victim but also disrupted the elephant herd's social behavior. Elephant herds are tightly knit matriarchal social units, with a lead female, her daughters, their daughters, and so on, as well as young of both sexes. The males are kicked out of the herd when they become teenagers—about fourteen years old, very similar to humans.

As part of my research project, several of us had developed a theory about how elephants could affect their own habitat. Because, next to people, elephants are the greatest animal land clearers—animal bulldoz-ers, capable of plowing up a terrain and knocking down the trees, de-foresting, and damaging soils—they could be what geologists called a "geomorphic agent," meaning that elephants could affect the very shape of the landscape. Once we became involved in trying to understand how changes in water and food supply affected elephants, and how elephants might affect their own water and food supply, we found once again that it was necessary to get answers to very simple questions: How much water does an elephant drink, and how much food does an elephant eat? Still innocent and naïve in spite of past experiences, I assumed that this information would be easy to find. Elephants were well studied, weren't they? There were excellent specialists in their behavior, led by my guide to Tsavo, Roger Atwood, whose mother-in-law had wandered through the wilderness of Africa with fifty porters years ago.

From Tsavo, I went to Zimbabwe, where I had heard there were ex-cellent counts of elephant populations as well as data on the health and reproductive rate of elephant populations. If I could only get the Zim-babwe Department of National Parks data about elephant populations, I could be home free; we'd have all that we needed to know to develop ways to forecast the future of elephant populations. I arrived one day at the Zimbabwe National Park headquarters, excited and impatient to see if I could obtain copies of the studies of elephant populations. For me it was a tense moment. Perhaps the parks department would not let go of its data. Many organizations are reluctant to share data about wildlife. They are afraid it will be misunderstood and misused; also, the scientists who worked hard to obtain the data want to publish it first and get credit for their work, naturally enough. If I could not get the data, we might fail in our project and not meet the terms of our research grant. Nor would we be able to help elephants and endangered species in the way we had hoped.

But my fears were empty. After a few minutes' discussion with some

of the park personnel, one of them went into a back office and returned
with a large box full of computer cards—the standard way of maintaining computer data at the time—which he handed to me. The box contained detailed information about four thousand elephants that had been
shot and then studied. For each one, there was an estimate of its age, its
health and vigor, and, for females, whether or not they were pregnant
and, if so, based on the size of the fetus, when they first conceived. This
was a unique set of data as far as I knew. I returned to the United States
believing that finally I would be able to make a definite contribution to
helping endangered species. But this information was only part of the information I needed. I also had to know how much water an average elephant drank and how much vegetation it ate each day.

Since elephants were so well studied, I assumed that for once the answers to these seemingly simple questions would be easy to find. But
again nobody seemed to know. We called the major zoos in the United
States, and the general response was "We just put the water out and
they drink it." In a few cases, the person we talked with would say, "Say,
that's a good idea. That would be a good thing to measure. We never
have. But we'll think about doing it."

Then I looked into how much food an elephant ate in a day. There
were several studies, some experimental, in which elephants were given
a weighed amount of food and the amount eaten calculated. But these results were surprisingly different, different by what scientists call an "order of magnitude"—ten times. There were only three studies that we
could find about how much food an elephant ate; they differed so much
that it was not possible to arrive at a reliable estimate of what they actually ate in a day. So I studied the scientific papers reporting what elephants ate. It turned out that the paper that gave the largest number was
done by a scientist with a very low budget, and all that he could buy that
could feed an elephant were slightly overripe oranges. He also chose to
measure the amount eaten by weighing the throughput—that is, the ele-

phant droppings. The overripe oranges gave the elephant diarrhea, so the scientist got a very large number. A messy research project in all aspects. So much for the science of elephant nutrition, at least when I was studying elephants.

During the flight over Tsavo, I asked Roger what had happened with his company that had sought to reduce elephant populations in a humane way. "I was doing this terrible, unpleasant job—killing elephants," he said. "I didn't like doing it, but I thought someone had to do it. I thought I was doing a good deed and that everybody would thank me for doing it. Instead, they hated me. Called me names. So, in the end, I gave it up. Right now I'm doing aquaculture—growing tilapia—about as innocent a fish as you can imagine—and trying to make a living at that."

I never learned how much food an elephant ate or how much water an elephant drank. But I learned something I had never expected: how hard it is to really do good and be appreciated for it, even when you have the best of intentions. Some days later, I got away from the noise of airplanes and Land Rovers and went off and watched a herd of elephants feeding. They paid me no attention, grazing steadily. Now and again they communicated with noises that sounded to me like stomach growlings— a low, rumbling sound. Some cut grass with their toenails, rhythmically swinging a foot as if it were a scythe, their foot making a soft, rhythmic, pleasant sound like someone raking a lawn. Others quietly pulled leaves and twigs from trees, now and again throwing dust over their bodies. The elephants seemed to be at peace and in control, neither trying to do good nor trying to do ill. At that moment, I envied their peaceful lives.

Twelve

HOW MANY LEAVES
ARE ON A TREE?

 arsimonious. That word went through my head as I looked around Tim Ericson's office. There was nothing on his desk except a telephone. Not a single sheet of paper was visible. Nothing decorated the gray walls. It was an inside office with no windows. The metal desk was the same gray color as the walls, lit by a fluorescent light in the ceiling. Tim sat at his desk, and took out a single pad of lined paper on which he wrote with a black ballpoint pen. Tim was a theoretical physicist at the ZAC (Zeon Aeronautic Corporation) Laboratory, where I had come to spend the summer. ZAC had invited me to work on what the corporation was calling a "socially relevant problem." The federal government was suing ZAC for preventing competition. As far as I could tell, ZAC had created a summer program for socially relevant uses of computers as a way to improve its public image. One of the topics the corporation had chosen was environment, and word had come to Yale University: Was there a way anyone involved with environment could think of using computers? I had been talking about my ideas for developing a computer program that could mimic the growth of trees; soon I found myself at ZAC's laboratory. I met with several ZAC scientists and mathematicians, and Tim was interested in the ideas I suggested for a computer program that grew trees in a forest. We soon were immersed in the project, but this was the first time I had been in Tim's office.

I had never seen a scientist's office like this. *Parsimonious* was the best word I could think of. Entirely different from my own office at Yale, with papers scattered everywhere, much like my office had been at Woods Hole Marine Biological Laboratory. A universe of difference from the office of my Yale colleague, Jim Jacaranda, with whom I had hiked in New Hampshire's Presidential Range. Jim collected everything and threw nothing away. His office had a floorful of boxes of computer cards, bundles of books, piles of papers—published and unpublished—and stacks of unboxed computer cards. Jim was a rock hound—a person whose hobby was studying geology out in the field and collecting rocks. Whenever he went on a hike, he carried a geologist's hammer, a small, narrow hammer especially made to chip samples from rocks.

Jim's collection of books, computer cards, and rocks occupied several shelves on one wall in his office. The shelves were the kind that were attached to metal strips on the wall, which were in turn screwed into the walls' studs. The shelves were attached with tiny hooks to holes in the metal brackets. His shelves were twelve inches deep.

I often admired Jim's rocks, fascinating and different from one another: a gray hard rock with small rectangular crystal-like shapes embedded in it; a granite chunk from the Presidential Range with a large band of white quartzite running through it; a yellow sandstone with horizontal bands; a conglomerate with pebbles the size of golf balls sticking out of its surface. Some were large—about a foot in diameter—and heavy.

To get to Jim's desk or to his collection of rocks, you had to weave your way through other piles of books, papers, and computer cards on the floor, through pathways that Jim kept just barely clear. When you came in to visit, he would graciously remove whatever he had piled on a wooden office chair set out for guests and motion you to it. There you and Jim sat, surrounded by as much of nature as he could squeeze into his office, along with much that had been written about nature. A great naturalist and expert on forests, Jim had entertained me with his wry sense of humor, as he did when he told me the story about the lady botanist and the cigarette chase down Mount Washington.

One day, Jim's collection of rocks got too heavy for the shelves, and the shelves gave way. Rocks were strewn across Jim's office, helter-skelter, with little rocklets that broke off during the fall scattered widely. I went by his office that day and looked in at the semi-chaos. Some shelves were hanging precariously, some had fallen to the floor, but Jim sat calmly at his desk working away, amid the litter of rocks and papers. I was sure the shelves would soon be repaired and the rocks reinstated in their rightful position on the shelves. But a month later only the shelves were fixed; Jim never bothered to move the rocks. Over the next months, I peeked in his office now and again, and everything was as it had been on the day of the rockslide, rocks still lying where they had fallen, boxes of computer cards still on the floor, a new path created around the rocks to Jim's desk, and Jim was sitting there calmly talking on the phone. As far as I can remember, the rocks were still on the floor when I left Yale and went to work at the Marine Biological Laboratory.

Jim's Yale office was in my mind as I stood in Tim Ericson's office, in shock.

"How do you keep your office so neat?" I asked.

"Only keep the most recent copy of anything," said Tim in his usual compact and terse way, his speech as parsimonious as his office. When he stepped out of the room briefly, I couldn't resist. I opened his desk drawers and his cabinets, sure there would be a hidden stash of messy papers. Nothing. Not even a *neat* stack of papers. The office was empty.

Essentia non sunt multiplicanda praeter necessitatem, William of Occam had written in the fourteenth century. "Entities are not to be multiplied beyond necessity." The phrase had become famous in modern science as "Occam's Razor," one of the fundamental principles of the scientific method. Translated into science, it means choose always the simplest explanation consistent with observations. It is the aesthetic scientific principle. Truth and beauty lie in simplicity. I knew the idea well, but I had never seen it manifested in a person's life. Tim Ericson certainly followed this principle in his lifestyle. His office was the essence of *Essentia non sunt multiplicanda praeter necessitatem.*

So was everything else about Tim. He engaged in no small talk. I felt we were friendly, and we certainly worked well together, but occasionally he and I would eat lunch alone at the laboratory's cafeteria, surrounded by tables of other scientists chatting away loudly—joking, laughing, teasing. Tim and I ate mostly in silence. Tim never filled any gaps of the silence during our lunch with small talk. No "How's your family?" or "Nice day, isn't it?" We would eat in silence unless Tim had something of value to say.

Sometimes the silence would bother me. I wanted to be friends with Tim, so I would say something just to be friendly. But if it was small talk, Tim wouldn't answer.

Tim also never made a mistake. His work was always perfect, excellent, nothing wasted. Computers were primitive by modern standards, and at the beginning of our work, our link with the computer was an electric typewriter. The computer would print out one line of the computer program at a time, unless you could specify a set of lines by their line numbers. This meant that you had to have in your head the structure of the entire program and, in detail, the contents of each line. Since there were hundreds of lines of computer code in the program, remembering all the locations was quite a mental feat.

The weaker-minded could get the entire program printed out and refer to it, but we were in a process of rapid development and those print-outs were out of date too fast.

I would have to think for a while and remember just where a line of code was and then try to type its revision. I could touch-type, but still I made typing mistakes and mistakes in logic. Often I had to refer to the printout of the latest version.

Tim never looked at the printed program and he never made any mistakes. He would know exactly what line of code he needed to fix, have the computer type it out, and then he'd fix it. Well, I exaggerate. By careful monitoring of Tim at work during the summer, I did catch him making one typing error. And I think I remember that once he made a mistake in his arithmetic or in his logic. Maybe. Or maybe I just imagined it—wishful thinking. *Essentia non sunt multiplicanda praeter necessitatem.* That was Tim.

He was a kind and gentle person. His hobby, I discovered about halfway through the summer work (hobbies were small talk, apparently, and not mentioned at lunch unless you knew and brought the subject up), was handcrafting children's wooden toys—the wonderful old-fashioned kind: wooden vehicles with wooden wheels that turned, wooden trucks, tractors, and automobiles. On weekends, Tim sold them at charity fairs. I forget how this information came out. I think one day one of the toys appeared on Tim's desk, in transit to a fair, and I asked about it. Tim never mentioned it again unless I asked him. Like his work with computer code, one parsimonious exchange about his hobby communicated all that he thought necessary. It was inefficient to repeat things.

Essentia non sunt multiplicanda praeter necessitatem. This was fundamental to the work we did. We put nothing in our computer program that was not necessary. And we proceeded as William of Occam had advised: Step by step. Try the simplest. If that worked, fine. If it did not, add the smallest possible complexity and try that.

Consistent with William of Occam's advice, we first wrote computer code that grew a tree. It did. Ericson said, "The computer calculates how many leaves are on a tree. Is it right? Are we even in the ballpark?"

As the working ecologist on the project, I said I would look into it and went back to Yale. Once again, this seemed a simple enough question. A child could ask it. Perhaps one of the children playing with one of Tim's handcrafted toys. I imagined a conversation between Tim and his son. (I didn't know if he had any children; that too was small talk, not relevant to our work.)

"Daddy, how many leaves are on a tree?"

"I don't know, but I'm working with an ecologist at a forestry school," Tim would say if he did have a son. "He'll know."

But I didn't know. After taking a peak into Jim Jacaranda's office and being relieved to see that all was well with the world—his rock collection still on the floor, the narrow path among various objects still there from door to desk—yes, some of us managed with human failings and frailties—I went to the Yale library.

Yale had then and still has one of the greatest libraries in the world. Its forestry school had its own extensive library. No Internet, no websites full of information, just books, books, books, papers, papers, papers. For two weeks, I spent my days in the forestry school library and wandering through dusty stacks in the main library. Books were everywhere, many old and rare, most neatly shelved, some just lying about, waiting for a librarian to refile them.

In the midst of the library stacks, I came across a book someone had left about and discovered it was a gazetteer of Cheshire County, New Hampshire, the very county where Heman Chase lived, where I had spent so much time in the woods. Brushing dust off of it, I discovered that it had been published in the mid-nineteenth century. I flipped open the pages and began to read. Soon I came across a statement that in one year the county was infested by "a plague of loathsome worms" that ate all the leaves off the trees. Ah. So nature suffered from outbreaks of insect attacks before Europeans introduced the gypsy moth. These were native "worms," actually caterpillars. They did not know about *Essentia non sunt multiplicanda praeter necessitatem.* They were messy and scattered things about. They were repetitive. I had walked in such woods and knew. The caterpillars in reference are what we call inchworms. They crawled on leaves above your head. They hung from silk-line threads so they could move from tree to tree, and so you continually walked into them. You could hear the rain of frass—the droppings or scat, the excre-ment of the caterpillars. Little scats hit you on the head and piled, un-tidily, on the forest floor. Dead leaves and parts of leaves from their untidy feeding habits lay on the ground everywhere. Wandering through one such partially eaten woodland near Yale, I considered my own situ-ation and the work I was doing at ZAC.

Almost two weeks had passed and I had not yet found a reference that gave the number of leaves that were on a tree. Embarrassing. A for-est ecologist at a forestry school who could not answer this question. Walking along, I suddenly remembered there had been one peculiar study done in the 1930s at Cornell University. A scientist had built a greenhouse around a small apple tree and measured the oxygen and car-bon dioxide going in and out—a way to measure the photosynthesis and therefore the growth of the tree. In the fall, he had cut down the tree and measured everything about it. I dug out that paper and discovered he had actually counted the number of leaves on the young apple tree. About ten thousand. Then in the forestry library, I discovered another paper that gave the count on a large and mature tree: About one hundred thousand. That was it. Other scientists had weighed the total amount of leaves on trees, but I could find no one who had taken the time to count all of them.

With this in hand, I went back to the ZAC Laboratory.

"Ten to a hundred thousand," I said to Ericson, avoiding all small talk.

"Right. We're in the ballpark," he said, and returned to typing on the electric typewriter connected to a computer. He was satisfied; we were hoping for precision and we didn't have it, but we had created a computer program that, in its simplest form, forecast what was known about the number of leaves on a tree. We had followed Occam's razor. Almost nobody bothered to count leaves on a tree—a tedious task, and a question that only came up when you really began to think hard about how a forest worked. We had to know, because we wanted to be sure, step by step, that we were doing enough but no more. Well, we could grow a tree that looked like a real tree and had the range of number of leaves that we could find information for. Now we could move to the next simplest step: growing a group of competing imaginary trees of the same species in the computer. *Essentia non sunt multiplicanda praeter necessitatem.* It worked for science.

But did it work best for scientists? There was something to be said for both Jim's and Tim's offices and their approach to life. Somehow there was a comfortable feeling in Jim's semi-chaotic office, much like a forest. But Tim's efficiency was undeniable and his mind as powerful as modern, well-executed science. *Essentia non sunt multiplicanda praeter necessitatem.* It described how to study nature, but it was not a particularly good characterization of nature—not with the plague of loathsome worms dropping frass and cutting away leaves in a forest.

Thirteen

HOW MANY BOWHEAD WHALES
EVER LIVED ON THE EARTH?

We struck that whale and the lines paid out
And she gave a flourish with her tail
And the boat capsized and we lost our darling boys
And we never caught that whale, brave boys,
Never caught that whale.

—FROM *GREENLAND FISHERIES,*
AN AMERICAN FOLKSONG

ne day when I was still working at the Marine Biological Laboratory in Woods Hole, I was eating lunch at the Fishmonger restaurant. The Fishmonger was a common gathering place housed in a weather-worn building just by the lift bridge that led into Eel Pond, the pond in the center of Woods Hole that served as a marina and was where the *Alvin*, the famous deep-sea diving vessel, was docked. You could just see the pond and the bridge out of the smudged, dirty windows of the restaurant. I sat at one of the rough-hewn, splintery wooden tables in the dimly lit room. The restaurant was full of its usual crowd, a mixture of students, graduate students, and research scientists, but few tourists. Dress was the Woods Hole usual—a reverse status dress code: grubby was good; grunge was anticipated here by several decades. The general attitude was that the more important you were, the sloppier you could and should look. My wife, Erene, who tried to dress nicely in a businesslike manner for her work at the laboratory, knew she was criticized for her careful and neat appearance. As she mentioned several times, people assumed she must be a secretary—only secretaries dressed as well.

Fall weather was turning and the damp coldness contrasted with the dry heat of the African plains I had just visited. The neat, starched, khaki uniforms of East African park staff and scientists—shorts and high socks—were replaced by aging blue jeans.

The food was pretty good and relatively cheap, and the Fishmonger was a popular spot with the scientists and students. I was musing about the many things that had nothing to do with science that kept interfering with attempts to apply science to wildlife, endangered species, to ecosystems, to life on the whole Earth. Superficial beliefs, going as far as how one looked and dressed, could often determine whether work was taken seriously.

I was trying to relax and enjoy a bowl of clam chowder when Fred, one of my research assistants, hurried in and sat down next to me, a little out of breath. "You've got a strange visitor," he said. "Nobody knows what to make of him but he insists on speaking with you."

"What's the problem?" I asked.

"He's in a three-piece pin-striped suit, is what," said Fred, "handkerchief in his pocket. Wearing those rimless glasses. Has a shiny leather briefcase, polished leather shoes. We think he must be a typewriter salesman, but then why would he want to see you? Anyway, he seemed determined and he's waiting at your office. We didn't know what to do. Some kind of freak," Fred said, "obviously a wimp, no way a marine biologist. He insisted on seeing you and not the purchasing officer."

"Doesn't sound so bad," I said.

"Remember, Dan, some of the old-time marine biologists already think you're a little weird, working at MBL and going to Africa to study elephants. Don't push your luck. Word to the wise," he said. "The word is out that nobody can understand what you're really doing here, what with computer whale games and experts on elephants coming to visit. Best to get this typewriter salesman out the door."

I hurriedly finished my clam chowder, bread, and herb tea and went back to my office. Just as Fred had described him, my visitor was elegantly dressed, a tall man who stood just inside my office looking around. I was suddenly conscious of my messy office, still cluttered with papers and still decorated only by the one lonely Christmas cactus surviving in a dingy corner on top of a bank of file cabinets. My office had the drab color of Tim Ericson's, combined with the confusion of Jim Jacaranda's, lacking only piles of rocks and boulders on the floor. This spiffy gentleman introduced himself.

"John Bockstoce, anthropologist, New Bedford, Massachusetts Whaling Museum." He handed me a business card and looked around for a place to sit. I hastily moved a pile of papers from an old wooden armchair and, dusting it quickly with them, offered him a seat. He carefully took off his elegant pin-striped suit coat and hung it on the back of the chair, first dusting the chair. As he did so, I noticed that large muscles bulged in his upper arms and torso, pushing against his starched white shirt. Once he settled into the seat, he began to talk.

"Been studying Eskimo culture—living with them," he said. "Learned their language. Got to know them well enough so they took me whaling with them. Soon I realized that to understand their culture, I had to understand Eskimo whaling. But one thing led to another, and once I got involved in trying to understand their whaling, I realized I had to do that against the background and history of Yankee whaling. So I started a study of the history of Yankee harvest of the bowhead whale—the favored whale of the Eskimo. I've been trying to reconstruct the entire history of that whaling." He shifted his long legs in the chair, feet pushing against a pile of my papers on the floor.

"Just was in Washington talking with the National Oceanographic and Atmospheric Administration. They said they'd fund my study if I could find a biologist who knew about counting animals and about endangered species. Several people recommended you. Get right to the point. I'm here to find out if you'd like to work with me on this project.

"We've got great historical records at the New Bedford Museum," Bockstoce continued. He explained that most of the great Yankee whaling ships made New Bedford their headquarters. "The Yankee whaling ships left New Bedford in the fall, sailed south along the Atlantic coast of South America, around Cape Horn during its summer, then sailed north in the Pacific, stopped at Hawaii to refit and take on fresh water and food, and then headed north to the Bering Straits and sometimes even farther north to hunt bowhead.

"All the whaling voyages ever made are known because newspapers of the day listed the departures and landings of every ship. And many of the logbooks from the bowhead whaling voyages still exist," he said. "We've got many at the museum. Know where most of the others are around the country. My idea is to use the logbooks to reconstruct the entire history of whaling—how many whales were killed, even estimate how many bowheads there used to be. That's why the NOAA people want me to find a biologist.

"So here I am," John said, wiping the dust from my office off his pant-leg and picking up a few pieces of paper from the floor as he talked to me, putting them neatly on the laboratory counter next to him. "So, would you like to work with me on this?" he asked.

We talked for several hours. John had an appealing skepticism and pleasing sense of humor. He had an honesty about what he did, about his experiences with the Eskimos, and about what he knew and did not know. It was a pleasant relief after my experiences with Richard Needles's tall tales about sperm whales acting like African gazelles. I expected that it would seem strange to my colleagues in ecology for me to start a project about bowhead whales with an anthropologist, but what the heck, I thought, I was working at Woods Hole studying elephants.

Because I had done the one study about the social behavior of sperm whales that had turned into a total disaster, I was especially attracted by John's description of what we might accomplish. Perhaps we could use historical records to find out something we could never learn any other way: how many animals there had once been and how their numbers had changed over time under the pressure of Yankee whaling.

"Sure," I said, liking John and the way he spoke about his work. He got up, muscles bulging under his shirt, put on his suitcoat, adjusted his suit's pocket handkerchief and his rimless glasses, shook hands, and headed out the door. I watched through my window as he passed by outside, and saw that he was the object of strange side-glances from Woods Hole scientists.

So we began an unusual cooperation: an anthropologist and an ecologist. John went around the country and located all the known existing logbooks. We began meeting regularly at Woods Hole, John always arriving in a three-piece suit looking as if he had just stepped out of an advertisement for Brooks Brothers. Occasionally we would eat lunch at the Fishmonger, where the expressions of the local scientists, seeing this elegantly dressed person, amused me. On one of these occasions, John told me that he would be stopping work temporarily on our project because he had to undergo extensive dental repair. I asked him what the problem was.

"A few years ago I spent about eighteen months in the Arctic, living with and studying the Eskimos, and generally traveling around on my own to explore old Eskimo sites," he said. "I was way out in the middle of nowhere, doing some studies of old Eskimo sites, and had to find my own food. Was out hunting for some caribou to eat, and I tripped on a patch of icy snow. Fell forward onto my rifle butt, knocked out all of my front teeth. Had to survive several months on my own that way, until I could get back to where there was transportation to the Lower Forty-Eight." My opinion of John, already high, rose even more. This was no wimp, but a person who could survive alone, wounded, in the Arctic.

After his teeth were repaired, John began his regular visits again. One day when we took a break for lunch, he told me that he had just returned from London from a meeting of the International Whaling Commission. "Had a curious experience," he said. "Some guy tried to mug me, right in a nice part of the city, but I shook him off—got the police to take him," he said simply, and then turned to his lunch and to a discussion of whales.

Our first task was to secure funding for our study. We wrote a proposal and I attended a meeting held by NOAA where I presented our ideas. Everything depended on the reaction of the other scientists at this meeting. I saw Richard Needles, the Canadian whale biologist who had led us astray with his tales of sperm whale behavior, and my heart sank. Here was a hostile audience. Then I spied Charley Russell, talking with his typical animation to a group of whale experts, dressed as usual in his casual once-white shirt open almost to his belly button. Before the meeting started, he and I talked, and I told him about the plan for me to work with Bockstoce.

"Bockstoce? He's amazing. Always dressed like a dandy, but don't underestimate him. He's one tough guy," Charley said. "You know, a few months ago I was standing at this bar in one of those gentlemen's clubs in London, one of those stuffy clubs, and there was this person standing next to me dressed in a three-piece suit. Only he was too young and too fit to be a member of the club. Decided he was a guest like me. We got talking and it was Bockstoce. He had a cut on his chin and I asked him about it. He said that someone had just tried to mug him, but he had fought the attacker off. I bought him a drink and while we were standing there, a policeman came in dragging this scruffy-looking fellow who was moaning and wailing," Charley continued.

"The policeman went over to Bockstoce and said, 'Excuse me, sir, but I believe you have broken this gentleman's jaw.' Very polite, the bobby was, and spoke very mildly, as if telling Bockstoce that he had just found him a better room or something.

"Bockstoce explained to the bobby that the man had tried to mug him."

This, I realized, was the story Bockstoce had mentioned to me in an understated sentence.

"He had hauled off and socked the mugger one, then calmly came into the club and ordered a drink. One tough cookie." Charley laughed. "Calm as a clam, sipping his drink. 'Broke this gentleman's jaw.' I'll tell you. If you work with him, just don't get him angry. Might break your jaw, too."

The debate over our proposal was intense and mostly hostile until Russell, having listened carefully to the methods we were going to use, got up and spoke, as if a lightbulb had just gone off in his head. He said to the others that he just now understood what we were trying to do, and even though the method wasn't perfect, it was the only one that would work, and we would do an important thing and ought to be funded. He went on to describe how the International Whaling Commission could use our information. That organization, a voluntary one of nations involved in whaling or interested in whaling, set catch limits for each species of whale. Charley explained that if we could find out how many bowheads there used to be, we might be able to help set a realistic limit on the catch allowed to the Eskimos. He carried the day and won the argument on our behalf. I tried to thank him, but he brushed it off and headed into town to find a bar and see if there were any good-looking women about.

Soon after, Bockstoce and I began our work in earnest. John collected logbooks from 20 percent of the voyages ever made. In those logbooks, the first mate wrote a report each day at noon about the previous and present day's activities. He wrote down the ship's location in latitude and longitude, sea conditions, the visibility in miles, and the percent of the ocean covered by ice, if any. He had a stamp in the shape of a whale and stamped that whale picture in the logbook margin every time a whale was caught. There was a blank in the center of the whale stamp and the first mate wrote within it the number of barrels of oil obtained from each whale. That number told us an approximate size of each whale.

We hired six people who spent six months typing into a computer the first mate's daily observations from all the logbooks Bockstoce had gathered. Then we began our analysis. We divided the northern Pacific Ocean into latitude and longitude zones, from Canada and Alaska to Siberia, and north into the Bering Straits and in the waters north of Canada.

With the computer, we were able to create maps that showed the numbers of bowheads caught in each part of the Pacific and the Bering Sea during each decade throughout the entire Yankee whaling period— from 1840 until just after World War I. In the early years of whaling, many whales were caught in the southern part of the bowhead's range— about halfway between Hawaii and the Bering Straits. But these whales were rapidly depleted and the catch moved northward, decade by decade. We learned from the data that a third of all the bowheads ever caught by the Yankee whalers were caught in the first decade—from 1840 to 1850—and two thirds in the first twenty years. During the next sixty years—from 1860 to 1920—the ships that left New Bedford for the arduous voyage around the Cape and north caught the remaining third. The catch declined rapidly, but even so the whaling ships continued to search for bowheads.

By the end of the era, voyages took three years. Ships sailed north of the Bering Straits, overwintered locked in the ice, and then hunted the whales when the waters opened and the whales returned. Often, a group of whaling ships would anchor near each other. A painting made by a member of one of these ships' crews shows men playing baseball and football on the ice in well-laid-out playing fields, cleared of snow, with whaling ships locked in the ice around them.

Officers sometimes brought their wives on the voyages. Food was plentiful from the sea and impressive banquets were held, for which a whaling ship would print formal invitations. One of these that still exists shows an elaborate menu of shellfish and finfish that would delight many a gathering today. It would be a rare treat, considering the over-fishing that has taken place since of most of the species listed on the menu.

Every year at Woods Hole, there was the Black Dog Contest, a kind of carnival that celebrated the intentionally casual character and dress of this research village. It was held on a large open green behind the labo-ratory, and featured high school students performing short plays, local scientists who had such hobbies as circus acrobatics, a man dressed as a clown who danced with a life-sized floppy, three-legged doll, and the central feature—the Black Dog Contest. Entries could be black dogs, but other kinds were accepted, including people dressed up as black dogs, people not dressed as black dogs, and anything else that someone thought was amusing.

I was wandering around the contest grounds, observing this celebra-tion of the new casualness, when Bockstoce turned up in a dark blue three-piece suit and matching tie. He said he was somewhat in a hurry, but had just come back from the annual meeting of the International Whaling Commission in Cambridge, England.

"You might think this stuff about bowhead whales and their catch is nothing—not important, but that's not how it's seen at the IWC," he said. "During the meeting, the Japanese delegation proposed an increase in their take of whales on the ground that this was a traditional cultural practice. The American delegation opposed it on the ground that any hunting of an endangered species was wrong. An Eskimo poured red ink all over one of the Japanese representatives, and there was danger of a fight," he said. "Then I overheard one of the Canadian delegates speaking to one of our American reps. He said, 'What are you going to do about your indigenous cultural problem?' The American looked confused. 'What cultural problem?' he said. 'Well, you just argued against the Japanese taking whales for cultural reasons, but you've got a treaty with the Eskimo that gives them the right to take whales. So you're contradicting yourself.'

"This made our study pretty important," continued Bockstoce, glancing at the current entry in the Black Dog Contest, a thin girl in black tights. "Suddenly, the Americans realized they had to negotiate an allowed harvest by the Eskimo, and needed to know how many that might be without harming the species. Someone mentioned our study, and now they want to know our results—how many bowheads had there been? So they can use this to estimate how many might be taken."

We were watching the clown dance with the three-legged floppy rag doll, about which Bockstoce made no comment. He shook my hand and hurried off, much to the amazement of the rest of the crowd that included many Harvard and Radcliffe students doing their best to look scruffy, most wearing fairly new blue jeans and looking somewhat uncomfortable. Here and there I could spot a student from New York City, because New Yorkers really knew how to be scruffy and there they were, in dirty T-shirts, worn jeans and basketball sneakers starting to come apart at the seams, looking quite at home.

So, I thought, here, finally, is a chance that we might make a difference in a real-world situation, actually help save an endangered species, and it was because for once we had good data. We were finally able to answer one of the simple questions that kept coming up in my work—how many bowhead whales ever lived on the Earth? And the key to it was, from the point of view of most of the other Woods Hole professionals, an oddball person who dressed all wrong.

I saw Erene walking around the Black Dog Contest by herself, so I introduced her to Bockstoce and asked her to join us. I was with the two best-dressed people in Woods Hole, I thought. John took his leave and Erene and I continued to walk and watch the other Woods Hole inhabitants. She invited my two children from a previous marriage and me for dinner that evening at her house. This was before I asked her to marry me, but the idea was working its way into my mind.

Our work took several years, and the method for estimating the original abundance of bowheads was not one of the strongest statistically. But it was the best we could do. In the end, our data indicated that there had been about twenty thousand bowheads at the start of the Yankee whaling era in 1840, give or take ten thousand—somewhere between ten thousand and thirty thousand whales. At the time the Black Dogs were contesting, there were about six thousand bowheads swimming in the Pacific.

The International Whaling Commission used our estimate of the original abundance and compared it with the present. The rationale was that the pre-Yankee harvesting number of bowheads was the natural number and an abundance that should have remained constant year after year as long as there was no interference from Europeans and Americans with their modern technologies. This led, through a series of political negotiations at IWC meetings, to a reduction by one bowhead per year in the take allowed to the Eskimos. Well, I had succeeded in finding out the answer to a question, and even more, a question about how many animals of a species there once were. Finally a success. But the effect on the conservation of endangered species was saving one whale a year. Not a world-shattering victory, I pondered as I made yet another attempt to clean up my office and put the papers from the bowhead study into some kind of order, sloppiness and order in sciences and in human life, my thoughts wandered to black dogs, old bluejeans, and the neatness of Erene and John.

My mind kept returning to one seemingly neat and orderly idea—the assumption of the IWC that there had been a natural number of bowheads that was unchanging—that it was natural just because no Yankee hunted them. This idea wore at me. I just could not accept it. The abundance of all animals had to vary with climatic change, changes in ocean currents, ups and downs in the production of food that an animal ate. Thinking it over, I decided that there was a much more straightforward approach. If there were six thousand bowheads now, and if we could give some kind of estimate of their reproductive rate, we could figure out how many whales might be harvested without damaging the population. One percent of six thousand is sixty. The IWC and the U.S. government treaty with the Eskimos was allowing a take of less than forty. Since the reproductive rate of large mammals was low, but not as low as one percent, this seemed to be a rather safe level of harvest—much less than the hundreds of bowhead that had been taken in some of the early years of the Yankee whaling. So the victory seemed less and less the more I thought about it.

But we had at least reconstructed a fascinating history of one of the world's great creatures, and we had created a set of maps like a moving picture show of how the bowheads had been hunted and how they had retreated northward, always northward, from the onslaught of the Yankee fishermen. Only the combination of what Bockstoce knew as an anthropologist and the kind of thing I knew about populations could have led to that reconstruction. We had not saved the world, but we had done a small thing that was unique. We had found a way to count a population of animals long dead and to track population changes over time, through written history. We had answered a seemingly simple question: How many bowhead whales used to swim the oceans?

And I had come to know one of the most fascinating characters I had ever met, one I have respected my entire life—John Bockstoce, one of the toughest and most rugged, but also one of the gentlest, kindest, and most thoughtful people I have ever known. I never talked much about him to the marine scientists at the Hole, nor to the people eating at the Fishmonger. Secretly, I hoped that they would go on thinking he was a typewriter salesman and that they would still be puzzling over what kind of research project the guy who was studying elephants using computers could be dreaming up to do with typewriters or with computers' new descendants, minicomputers.

Fourteen

☙

TRIMMING ELM TREES

ob Nickel and Harry Chains were short of money. They were students at the Yale School of Forestry, and their courses had convinced them that they knew enough about forestry to act as professionals. They went into the business of tree-trimming. They were vigorous, outdoorsy young men, hardworking, earnest, and with a good sense of humor. Bob was about six feet tall, with black hair, a thick black beard, and a pleasing manner. Harry was a wiry Australian, about 5-foot-10, with a narrow face and a determined manner.

Bob and Harry had worked for me on research projects in the forests of New Hampshire. By this time I had figured out, I thought, a way to combine the two worlds I had been living in—the nineteenth-century self-contained world of Alstead, New Hampshire, and the increasingly hi-tech world of modern science. The computer model that grew trees had to be compared with real data. I spent one summer getting some of that information and hired Bob and Harry to do the main fieldwork. We worked in the White Mountains of New Hampshire, where Jim Jacaranda had followed the cigarette-smoking woman botanist to the summit of Mount Washington. I got U.S. Geological Survey maps of the mountains, making use of skills I had learned from Heman Chase about map-reading as well as map-making. And on these maps I selected points at random, so there could be no bias in what we chose. Bob, Harry, and I then went to those plots and made measurements that I hoped we could use to test the forecasts of the computer program. We found our way with a map and compass, doing what later became known as "orienteering," an enjoyable outdoor challenge. With this work I was able to hike through the forests and study them, following the ideas that had developed when I had helped Heman Chase survey the old farms in southern New Hampshire.

During that summer, I came to know Bob and Harry well. Bob brought along his dog, a large retriever whom he named "Henri," pronouncing it in French as a joke since that made it almost impossible to shout. Once we climbed Mount Moosilauke on an overnight trip, making measurements on the way up and the way down. Henri was impatient and always eager to go ahead of us, returning as fast as he could run and dashing past us, then turning around, visiting with us for a short while as if to say, "What's taking you so long?" Then he would head up the mountain again. By the end of the day, we reached the summit and cooked our dinner over a camping stove. Bob realized he had forgotten any food for his dog, who stood and watched us eat. Bob fed him some of the beef Stroganoff that came in a dried form that we had mixed with water, and he gave the dog water to drink. Most of the time Bob was completely competent and well organized, but once in a while, as in this situation, he was like most of us and would forget to bring something. We petted Henri and tried to make him as happy as possible on his limited food. He did not seem to mind too much.

Bob and Harry had plenty of vigor and could outhike me anytime. With their enthusiasm and good humor, I thought they could succeed at almost anything they wanted to do. But then their reach exceeded their abilities. Their first job as tree-trimmers back in New Haven in the fall was to trim a big elm tree near the curb on a busy New Haven street. The tree belonged to the owners of a gracious white-clapboard home, a classic New England house except that the front of the house was dominated by an ostentatious porch with Greek-style columns holding up the overhanging roof.

Bob and Harry arrived with a few saws and axes and some long ropes. The American elm was distinctively graceful and was once the dominant street tree in many towns and cities. Because of Dutch elm disease, few American elms remain today and their graceful shape is now unfamiliar. Typically, an elm's main trunk splits twenty to thirty feet above the ground into two to four arching limbs that spread out in what may be referred to as a vaselike shape. The result is a beautiful tree whose leaves, hanging down from the high arched limbs, shade the street below without limbs and branches sticking out into the traffic. These limbs were thick and strong, usually more than a foot in diameter. Elm wood is extremely tough, in part because the grain spirals upward, making the trees difficult to split with an ax and wedge, and difficult to cut when the wood is green.

Bob and Harry's elm was probably beginning to suffer from Dutch elm disease. One of the limbs hanging out over the street was dying. It was this limb that posed a hazard, and the homeowner had asked Bob and Harry to cut it. They started in earnest, one of them climbing up the tree and the other putting ropes on the limb. They were trying to keep the limb from falling onto the street—possibly onto a passing car. Their plan was to tie the rope near to the end of the limb and loop the rope around the trunk of the tree, using the loop as a brake. One of them would hold onto the end of the rope, pulling it against the loop. Friction would stop the limb from falling. Then they could ease the limb over to the lawn.

The limb was much heavier and stronger than they had guessed. As Bob sawed through the base of the limb, Harry held on to the end of the rope, standing near the porch to keep as far from the limb as he could. When the limb began to bend, Harry realized that the loop they had made around the tree trunk was not enough to hold back the weight. The rope began to pull him to the tree. Holding on to the rope, he rushed up on the porch and looped the end of the rope around one of the porch columns, expecting this to be made of solid wood and capable of holding the weight of the limb. But the column was a hollow façade. As Harry pulled on the end of the rope, now twisted around the column, the column groaned and started to rise into the air, the weight of the elm limb pulling it out of its base. Harry rushed closer to the column. He pulled with all his might, but the limb was stronger than he was and weighed a lot more. In a moment, Bob, watching from the tree, saw the column and Harry pulled up into the air, neither one grounded. Only the overhanging roof was preventing the limb from crashing into the street and yanking the column and Harry in a flying arc toward the street. Bob wasn't sure what to do first. If the limb fell, it was in danger of damaging a passing or parked car, possibly causing serious injury to passengers. The rope was threatening to destroy the column and possibly the entire porch, and perhaps injure Harry.

A crowd gathered to watch Harry's acrobatics with wonder and amusement. For a few more moments, Harry and the column swung like pendulums, suspended in the air by a rope. Bob climbed down from the tree and joined Harry to help pull on the rope. Their combined weight was just enough to bring the column back down to the ground and keep the limb in the air. But what to do next? If Bob let go, Harry would spring back into the air. But Bob couldn't stay on the porch forever.

Eventually, a few men in the crowd joined in to help. One took Bob's place holding the rope and another helped Bob finish cutting through the limb. As the crowd cheered, Bob pushed the cut limb, still suspended by the rope, away from the street. Harry and the helpful stranger slowly eased off the rope, lowering the limb to the lawn.

So ended Bob and Harry's career as tree experts. It was a lesson in homilies: the limits of academic education; the danger of a little knowl-edge; the limits of the enthusiasm of youth; the value of experience. When we seek to find practical solutions to the conservation of our forests, and when we do this from a distance, with little learning and less experience, we should remember Bob and Harry trimming the elm tree in New Haven. It's not a bad thing to think about in confronting life in general.

Fifteen

THE ECOLOGY
OF SPLITTING WOOD

nother way that the two worlds and two centuries I lived in came together was the ecology that the forests of southern New Hampshire taught me. The forests and the trees showed me many things that helped me as a scientist to understand my chosen science. During the time that I lived in New Hampshire, I learned a lot of what I know about forests and trees from splitting wood, cutting down trees, cutting the trees into logs short enough to fit into a woodstove or fireplace, throwing the split logs into a woodpile, and then putting the wood into the fire. When you split a New Hampshire red oak, there is a funny smell, a little acrid, a little rotten, almost like a baby's puke. Split sugar maple just smells nice—fresh, maybe a little sweet. White pine smells of pitch. So do hemlock and spruce. Usually, the pitch gets on your hands.

American white ash—a nice tree in general—is the easiest to split. Its grain is straight. That's why it makes such good baseball bats and handles for many tools. The grain is so straight and the wood so tightly strung together that an ash log sometimes literally jumps apart when you strike it with an ax. It makes a nice *pong* and the split pieces fly in all directions. You don't want to stand too close to somebody splitting white ash.

The hardest tree to split is American elm—or used to be. That tree has been on its way out for many decades because of what has been called Dutch elm disease, a fungal disease introduced from Asia that arrived via a shipload of logs from Europe, so the story goes. Elm is hard to split because the grain twists up the tree. You can't make nice boards from elm. The wood grows in a kind of corkscrewing pattern from the bottom of the tree to the top. The wood clings to itself. When I was living in Acworth, New Hampshire, a beautiful elm in the backyard succumbed to Dutch elm disease, and I took it upon myself to cut it up and split it so that at least in its death the tree did not go to waste. My rented house was heated only with firewood, which I either cut myself or paid somebody to cut. At the time, before I got a doctorate in biology and became a professional biologist, I was an unemployed writer with a wife and small daughter, and paying for wood wasn't possible. Cutting wood into chunks—short logs about two feet long, a good length for firewood—was rather easy, even with elm, and especially with a power saw. But the logs were big—too big to fit into a fireplace or woodstove—and it was necessary to split them.

The way you split a chunk that is difficult is to use a combination of axes, wedges, and sledgehammers. First, you try to drive a wedge into one end of the chunk. If the chunk is ash, as I said, the wedge will go right in and the wood will jump apart. But if it is elm, the wedge goes deeper and deeper. The corkscrewing grain expands a little, but like a green alligator, bites down on the wedge and grabs it. The harder you hit the wedge with the blunt end of an ax, the harder the wood bites back and holds on. Eventually, the wedge gets stuck and you can't get it out. So you get another wedge and drive it into one of the little cracks that the wood did allow to open, just a bit, because of the first wedge. If you're lucky, the first wedge will fall out and the second will get stuck; the net result is that the chunk is a little more split apart. But just as of-ten, the elm bites back on both wedges. Many's the time I have had four wedges stuck in the same chunk. Usually when that happens, it's time for lunch, and so I walk away, allowing the elm wood to squeeze back and tighten itself even more on the four wedges.

Taking a break helps, because then one can analyze the situation and try to figure out how to outsmart the elm wood. It's a matter of driving a wedge in at just the right angle or getting lucky with a hard blow with the ax. The problem about going directly to an ax is that the elm can also bite back on it and hold it, so that you have four wedges and an ax stuck in a two-foot length of elm. When this happens, there are several choices. One that I like is to put the whole thing in a big fireplace and burn it. The wedges and ax head come free and all you have to do is get a new handle for your ax. You get the definite feeling you have beaten the elm and at a small cost. But you need to have a large fireplace for this solution.

Another solution is to get a sledgehammer and keep hitting at the wedges and the ax until something gives. Often, the something is just a wedge, which tumbles out and falls to the ground, leaving three wedges and the ax still stuck in the chunk. The wood then closes up where the wedge had been.

Another solution is to slowly remove each wedge and the ax by any means possible, put the chunk aside, and come back in two years. By that time, the natural processes of decay from fungus and bacteria, helped a little by insects and other small animals without backbones—worms of one kind or another, I was never very good at identifying them—soften the wood. It becomes punky. It's not much good as fuel. Although the ax and the wedges go right in, they tend to go in as if the chunk were a tar baby. Nothing much happens. The wood doesn't split, it just gives. You can pull the wedges out, but you don't get a nice chunk, and if you do get a chunk at all, it burns poorly, giving off little heat and lasting only a short time. Doing this, you learn a little about the ecosystem of forests and the importance of the process of decomposition, as well as becoming acquainted with some brightly colored fungi that you might never notice otherwise.

Eventually, I got most of the elm tree split and it made pretty good firewood: not as good as sugar maple but better than pine.

It takes a lot of imagination or a big vocabulary to split an elm tree. Otherwise, you quickly run out of four-lettter words and other ways to describe the elm. The two-foot chunk, that little piece of nature, becomes a hated enemy who seems capable of many devious devices to foil a simple human task. I guess I don't recommend splitting elm wood to the faint-hearted or the nonverbal.

Trying to split elm helped me understand why it made such a beautiful tree. People aren't familiar with American elms much anymore, but they used to line many a street and many a college quadrangle as I mentioned before. They created a beautiful arching shape of huge limbs. The wood had to be very strong to hold up the big limbs at such elegant angles. It was that corkscrewing growth pattern that made it possible, I decided.

I have long wanted to organize a wood-splitting contest with my own rules. There are a number of people whom I would like to involve in this contest: people generally known as charming and gracious, but who, as scientists, have stolen my ideas and used them without giving me credit; people who have screwed me out of research grants so they could get the money instead or those who have scribbled reviews of articles I have sent to professional journals, writing on half a piece of paper from a yellow pad a series of invectives about me and why nobody should publish my paper—with no discussion about the scientific content or quality of the paper.

The contest would be to see who was better at splitting wood. My opponent would have to be rather unfamiliar with working with logs. I, of course, would be quite familiar with them. (After a few years of splitting and throwing wood chunks, I had learned to identify trees from their bark. In fact, there was a time that I was better at identifying trees from their bark than from their flowers.)

I would select two wood chunks of the same diameter, casually giving my opponent elm and myself ash. I would go first and with a simple *pong* blast the chunk into neat firewood. Then my opponent, not knowing the ecology of splitting wood, would try the same thing with elm. Depending on his fortitude, we could let him go on until nightfall. We would have lunch breaks and other entertainment while he buried one wedge after another in the elm.

Sugar maple is far and away the best firewood in New England and it splits rather easily too, because its grain, though very dense, is generally straight. You can make a fire of two or three well-dried sugar maple chunks, bank them carefully and go to bed, and you only have to get up once in the middle of the night to keep the fire going and the pipes from freezing if firewood is the only way your house is heated. By comparison, white pine, like most conifer softwoods, is full of quickly burning pitch and is a lighter wood that burns rapidly. You would get little sleep trying to keep a house warm with only white pine.

Trees that grow fast tend to burn fast. These are trees that you find in young forests. They grow well in the bright light of the open area, but that fast growth gives them little structural strength and little staying power as fuel. Sugar maple and red oaks, found in older forests, grow much more slowly. Their beautiful fine grain is one result and so is the dense wood that burns well but slowly.

For insects, oaks are the Baskin-Robbins and Ben and Jerry's of trees. Insects love oaks. Fungi do too. Oaks have hundreds of insect parasites. Some make galls on the leaves. They burrow into the bark and eat at the wood. They eat the leaves in many different ways. Leaf miners are among the most curious. They are tiny insects that live and eat inside leaves, feeding on the tasty green part, leaving the nonnutritious outer parts of the leaf, so the result is a translucent skeleton of the leaf— formed but not functional. You notice the great attention insects and fungi pay to oaks while you split an oak chunk and throw it on the pile. Sometimes when I was taking a break from a particularly nasty elm, I would examine a piece of split oak and admire the handiwork of the many insects at play inside.

Oaks fight back. They produce what are called "secondary compounds"—in this case, pesticides that are not primary to the metabolism of the tree, but kill or deter insects and fungi. I have always assumed that the smell of a just-split oak is in part from some of these compounds and in part the smell of fungal decay. In any case, I became familiar with the micro-ecological system of insects, fungi, and oaks, watching nature in action at a small scale. This was part of my education in the ecology of trees and forests.

Learning the ecology of splitting wood is a vanishing skill because somebody invented a hydraulic wood splitter that can put the power of a Cat-D8 bulldozer against elm and because few people depend totally on wood to heat their houses. My father-in-law Heman Chase, with whom I spent many a day splitting wood, invented a mechanical wood splitter. It was a cone-shaped screw device that bored its way into the wood in an ever-widening spiral. It worked and he patented it, but you had to have a tractor to run it, and it was big and clumsy. It lost out to the hydraulic thing. I guess after thirty or forty years, Heman got kind of tired of splitting wood, ecology or no ecology.

Heman heated his house solely with firewood for thirty years. He had converted a coal-burning furnace to burn wood. Otherwise, the heating system was a standard modern one of the time—central heating with hot water in radiators. He had a fireplace in the living room, for comfort and enjoyment as much as for warmth, and a small woodstove in his office down in the basement. Ever the country surveyor and a lover of numbers, he kept track of what he used and told me that on average he burned ten cords of wood a year. That's ten piles of wood each 4-by-4-by-8 feet—a lot of wood! By comparison, Missouri River steamboats burned ten or twenty cords a day, and usually either blew up or sank after one or two voyages.

On his more than two hundred acres of land, Heman kept sixty acres as a woodlot. Much of his firewood came from that lot, but he also removed neighbors' trees that died and fell over their driveways, or had blown over in storms, and he took occasional trees from other parts of his lands. I found it another ecology lesson to wander around the woodlot. This was sustainable forestry in practice before the term became popular. It was a woods, all right, but to the practiced eye quite different from the other woods Heman owned. He harvested most species in the woodlot except poplar (trembling aspen, *Populus tremuloides,* to speak scientifically). "It's a trash tree," he said, "good for nothing." It was one of the earliest of early successional trees and grew very fast. As a result, its wood was soft and burned quickly, not worth the effort to cut when there were oak and sugar maple. It was interesting to see, on visits I made later to Europe, that much of the flatland plantations in Italy and France were European or American poplar, so deforested was that landscape and so desperate were the people for any kind of wood. Poplar makes decent shipping boxes, but not much else, at least from a New England perspective.

Heman never had to plant trees in his woodlot. One of the blessings of New England's forests is how richly reproductive they are. Seeds and seedlings come in rapidly, spread by wind and animals, and the process of forest succession happens naturally and rapidly. This was enjoyable to see, nature busily at work regenerating her forests. Not every area where timber is cut is so fortunate.

The woodlot also was more open and had generally shorter, younger trees. If you were not familiar with the ecology of splitting wood, you probably would not have noticed. It was a rather subtle difference, and the surrounding woods varied a lot anyway. But after several years of logging and splitting wood, I found the sustainable woodlot intriguingly different. It persisted; it was just different. It did have the feel of a weed lot—densely growing, closely packed, young trees, heavily poplar, with much better firewood trees coming in here and there. It was a pleasant enough area that I considered building a small summer cottage on it. I made a clearing for the cottage but never got it built. I think elm trees took too much time so I could never get to that cottage.

Splitting wood for the fire, one's mind wanders to speculations: Why is sugar maple wood dense and white pine less so? It has to do with their ecological roles, their niches, and how they fit into the process of the re-covery of a forest—forest succession. That's the kind of ecology knowl-edge that comes home as the blisters and calluses build while splitting wood. It is the reason there is so much to the ecology of this activity. I recommend it to all who love nature and want to defend the environ-ment. The difficulty is how to plan a future that has enough forests and not too many people, so that whoever would like to could share in this educational activity and, as a sideline, expand his vocabulary.

Sixteen

POLITENESS UNDER GUNFIRE

hots rang out and I could hear bullets rattling around in the branches of the beech tree just above my head. I was on the eastern edge of Hutcheson Memorial Forest, the last remaining uncut forest in the state of New Jersey, taking a few people on a nature walk through the woods, part of my job as caretaker. It was a warm fall afternoon, one of those Indian summer October days that seem to be summer stretched out to dry. Some leaves were turning color but many were still green. There was a musty scent from the fallen leaves on the ground, and the blue sky had a few hazy clouds.

I had come across people with guns in this no-trespassing, no-hunting preserve. Part of my job was to patrol it and keep it a preserve, undisturbed by trespassers. But I had never actually been shot at. My initial reaction wasn't fear; it was anger. I was just plain mad that anybody would be shooting into this last stand of virgin forests, especially with me in the line of fire. Taking cover behind the tree, I waited for some more shots to see where the gunfire was coming from. The next volley made it clear that someone was shooting from the land next to the forest, just east, where there was a private house. I guided the guests quickly away from the line of fire and took them back to the visitors' parking area. I made sure they were safely on their way and then got on the phone and called Murray Buell, the director of the nature preserve and my major professor.

Murray, a great naturalist, was also a skillful fund-raiser and spokesperson for nature. For years, he had studied this small uncut woodlot near Rutgers University, where he was on the faculty. The 65-acre woodlot had been owned since 1701 by one family, the Mettlers, the original Dutch settlers. A few years before my arrival at Rutgers, the Mettlers had decided to sell the woods and Murray had succeeded in raising funds to purchase it as a nature preserve for the university. He had raised money from Sinclair Oil and the carpenters' union, among other sources, and agreed to name the woods after one of the heads of the union, Hutcheson.

Murray had persuaded the carpenters' union that the woods needed watching and therefore needed a live-in supervisor, and that the carpenters should build a house for this caretaker. They did, and Murray turned the caretaker's job into a scholarship for a graduate student. I was fortunate to obtain that scholarship, and my young family with my two toddler kids was able to live near the beautiful forest, with a stream running out from it down near the house. I gave nature tours on weekends and patrolled the woods frequently, preventing people as much as I could from removing plants and shooting deer in the forest.

Not too many days before the shots rattled over my head, I had confronted a hunter illegally dragging a deer out of Hutcheson forest across one of the research areas—a series of previously farmed fields that were growing back to forest, and were in use in experiments as comparisons with the ancient forest. The hunter dragged his deer right through

Seventeen

IS IT OKAY TO LET YOUR DOG
DRINK FROM THE TOILET?

ollution concerns most of us, but we have differing atti-
tudes about what is clean. In New Hampshire, there was
quite a variety of approaches to cleanliness, food, and disposal of wastes. My mother-in-law, Edith Chase, said
there was a family who lived in the backwoods who ate dinner at a large
table that had drawers in the sides. When everyone finished dinner,
each person wiped his plate clean and, without water or soap touching
it, put the plate into the drawer, to be ready for the next meal. Edith
swore that this was true and she was a Quaker, so it must have been.

Edith also told me another story about an upper-class family from
Concord, one of the bigger cities of New Hampshire, who had bought
a summer house in Alstead. They were invited to dinner at one of the
locals' homes not long after they arrived. When they got back from dinner,
one of their friends asked what it was like. "They was livin' just like pigs,"
said the missus. "They had the milk bottle right up on the table."

Then there was Heman's sister, Mary, who lived with her husband, Walter Burroughs, the gentlemanly alcoholic, in a grand brick house—the original house that went with the farm—just a short way uphill from the old mill. Mary and Walter had many cats, which they let run free around the house. When we ate dinner at Mary and Walter's house, the cats joined us. When Walter finished eating, he called to the cats and several of them jumped up on his shoulders and then down onto the table and ate the leavings off his plate, while the rest of us kept up polite dinner conversation.

In the winter, the cats had the run of the attic, which they used for every purpose. Once, Mary invited me to look at some of the old furniture she had inherited, which she stored in the attic, and asked if I would like any of it. There was a large wooden filing cabinet of an elegant nineteenth-century design, with drawers of many sizes. But the wood was discolored to a dull gray, so much so that I could not tell what kind of wood it was made of. The color was the result of generations of cats climbing over the filing cabinet and urinating on it, I suppose to mark it as their territory. I accepted the cabinet gladly and made a winter project of hand-sanding it, during which I discovered it was a made of beautiful oak. I finished it with old-fashioned varnish, not polyurethane, because the varnish enhanced the golden oak glow of the wood. At least the effects of the urine of cats was reversible.

Most families had dogs, and dogs being dogs, most of them drank water out of the toilet. Cats did too. Some cats and dogs seemed to prefer toilet water to a pan of water by their food plate. People wondered, as I find people do just about everywhere, if it is really okay for a dog to drink out of the toilet. Will they get sick or pick up some parasite that they could spread to people?

Later, I became friends with David White, who had gotten both medical and doctoral degrees at Rockefeller University in New York. Dave's doctorate was in microbiology and he did research on microbes that live in the soil—an intricate community of many tiny creatures that play a largely unknown but very important role in our lives and in keeping all life going on the earth. It's not an exaggeration to say that, for some important ecological processes, these microbes are necessary—that we completely depend on them—and losing them would make a lot more difference to our lives than the loss of some of the big, warm, and fuzzy animals that appear on posters about conserving nature. They are more important even than dogs.

It's also true that these almost invisible creatures are very hardy; it's unlikely we could do them in. Some of them add nitrogen to the soil, making it fertile. Others decompose all kinds of wastes and dangerous chemicals. Relatives of these microbes live in ponds and streams and are also important to life there. And then there are their nasty relatives that cause diseases. Some are good guy–bad guy bacteria, benign enough or even helpful most of the time, but if the environment for them goes bad or changes in certain ways, they take advantage of the situation and cause us and other animals serious problems. Ecologists call them "facultative" or "opportunistic" organisms because they change what they do depending on what the system offers—a kind of entrepreneurial microbe, investing biochemically in what is advantageous in the environmental marketplace at the time.

David White practiced medicine but spent most of his time on his ecological research. One of his hobbies was collecting ducks from all over the world. For a while, he was on the faculty of the University of Florida and told me that he lived on a street known as "physicians' row," full of trophy houses whose owners were proud of them and their clean neighborhood. Dave said his ducks made him pretty unpopular there, because 120 pet ducks leave a lot of droppings, some of which cast a scent through the neighborhood that made it seem as if the town sewage treatment plant was at the end of the block.

A few years ago, Dave and I were attending a scientific conference in

Washington, D.C., and we went out to dinner together. I asked him what kinds of studies he had been doing recently—the usual question one scientist asks another—and he said he had just completed a study to find out if it was okay for your dog to drink from the toilet. Dave had a wonderful, whimsical sense of humor and was able to discuss the most arcane knowledge about biochemistry, soils, and tiny creatures in the soil in a charming way that anybody could understand. His sense of humor always surprised people because he wore dark, horn-rimmed glasses that gave him a serious, academic demeanor. I thought this was one of his whimsical stories and, in a sense, it was, but he said, "This is a more serious problem than it may appear. Especially as in many areas of the world, fresh potable water is becoming scarce."

What set him on this scientific research project about dogs and toilet water, he said, was a growing number of proposals to have dual water systems. One system would have potable water and the other would have "gray water"—water that had been through, say, your washing machine or dishwasher and then went through some kind of filter and could be used to do such things as flush the toilet and water the lawn.

"A hue and cry erupted," Dave said, "when dog and cat lovers heard about these proposals. They started to worry about the health of their pets, who regularly drank from the ever-convenient toilet bowl." So Dave, the duck collector and expert on what happens in soils and waters, began his study of whether or not it was safe for your dog to drink out of the toilet.

He said that no matter how well and often you clean your toilet, there are bacteria in the environment that stay there and produce a thin layer of an oil-like substance that floats on the surface. It's one molecule thick and known as a "biofilm." These films have recently generated a lot of interest—there may be many practical industrial applications of such thin films.

"It turns out that there are two major kinds of bacteria that make these films in your toilet," he said. "One kind is pond bacteria and the other is fecal bacteria." The pond type are generally benign or at worst opportunistic pathogens. They only become a problem when certain uncommon situations force them out of their usual habits. Then there are "the really bad guys like most serious human pathogens—plague, enteric pathogens like salmonella and *Escherichia coli,*" he said. Then the going got technical because Dave really knew the biochemistry and the microbiology.

"These are Gram-negative bacteria," he said, "and these kinds of bacteria produce a fatty acid that forms the biofilm on the surface of the toilet water. The good guy—pond bacteria—and the bad guy—fecal bacteria—both produce these films. But they differ chemically, and you can tell by analyzing the chemical composition of the biofilm if it was made by good-guy or bad-guy bacteria. Even the most fastidious toilet bowls cleaned daily contain a biofilm at the water-air interface. You can't feel it or see it, but it's there."

Dave got a research contract from the National Water Research Foundation to find out whether it was the good guys or the bad guys who were making the toilet bowl film. He took samples from many toilets, using standard scientific sampling procedures. He found that toilet biofilms were almost always made by the good-guy pond bacteria.

Dave also looked at the difference between high-flush and low-flush toilets—low flush being the more environmentally popular one because they save water. "It didn't make much difference except that the higher the flush rate, the healthier the bacteria in the biofilm," Dave said, in his usual understated, whimsical way.

So, he concluded, "It's safe for your dog to drink from the toilet. And the more water you flush, the better off the bacteria. So if you're a bacteria buff like me, you'd want a high-flush toilet."

As usual, Dave had come through. He had actually succeeded in answering an environmental question, unlike so many other ecological projects I had participated in over the years. He really knew his stuff, did solid research, had his question clearly in mind, and did the measurements carefully and accurately. And he got an answer to a question that many people wanted to know.

So the next time you see your dog lapping out of the toilet bowl, you can thank M.D., Ph.D. David White, one of the best and most imaginative ecologists I have ever met, with a sense of humor about his work that is unequaled, and a whimsical way of explaining the most complicated chemistry and biology so that even a regular ecologist could understand it.

This leaves open the question of whether it's okay to have the milk bottle right up on the table or your cat cleaning your plate at the dinner table, but there is always time for another research project. That's one of the fascinating things about ecology.[4]

Eighteen

WINDS OF A
CONDOR'S WINGS

 aving looked long enough out the window at Woods
Hole and its boats and ships, caring for my lonely Christmas cactus among the piles of papers and books in my
dark office at the Marine Biological Laboratory, I decided
it was time to leave and to move west, out closer to where I was born
and where I had always wanted to go. I took a job at the University of
California in Santa Barbara.

The beautiful California coast and mountains behind the city and the
bright, cloudless days seemed to promise a new beginning. Perhaps here
I could really solve an environmental problem. And not long after I
moved to California, an opportunity arose to try to help once again to
save a small piece of nature. I was asked to join a scientific committee to
advise the state of California about how to save the condor. It was 1980
and there were still twenty-two condors in the wild, the birds with the
largest wingspan of any species in America, soaring across the dry

grassy hills, sandstone cliffs, and scruffy dry woodlands of Los Padres National Forest fifty miles inland from Santa Barbara. Condor numbers had been dropping for decades—throughout the twentieth century—and the number twenty-two was setting off alarm bells among environmental organizations, ornithologists, and naturalists across the country. The state wanted to do something, and formed a small panel of five experts to advise it on saving the condor from extinction.

Now this was an interesting situation, because at that time I knew little more about condors than I have just told you and had never seen one flying. In fact I had never seen a live one, not even in a zoo, or even a stuffed one in a museum. Some kind of expert, you might say.

We have pretty fixed images about experts, as the people who make television advertisements know. I had already experienced the downfall of believing that a recognized expert on sperm whales' social behavior actually knew anything about it. On the other hand, John Bockstoce, the anthropologist who studied Eskimos and whaling, was a true expert, not just about bowhead whale history, but about self-defense in the city and survival in the wilderness.

Part of the reason I had chosen Santa Barbara was the advice of Murray Buell, my major professor, the quiet, soft-spoken native of rural Massachusetts. He had spent a semester at the University of California in Santa Barbara; when he came back, he visited me and said, "Dan, you ought to take a job there."

"Why?" I asked.

"You'll just like it," he said and smiled. There was no doubt that Murray was an expert about ecology, and he was famous as an advisor to graduate students, another expertise. I trusted his judgment about my professional and personal life. Murray fit the standard mold of an expert—a university professor, past president of the Ecological Society of America, a gentleman who usually wore a tie, stood straight, and spoke with humility but great knowledge. It was hard not to take Murray for anything other than an expert, unless you happened on him in the woods in his field gear, an old floppy hat on his head, a small backpack slung casually on his shoulder, dust and soil on his shoes. There you would find him walking leisurely or looking through a tiny hand lens at some leaf, twig, or flower, or standing back, hands in his pockets, watching a tree as if it were going to do something interesting and unexpected. Murray should have been walking around in the forest in a white laboratory coat if he wanted to look like a conventional expert in those situations.

Although I had lived most of my life in the east, I had been born in Oklahoma and somehow always thought of myself, down deep, as a westerner. I expected that someday I would live out west. California seemed to be my destiny. And I decided Murray was an expert about me, as well, and must be right in his intuition that Santa Barbara was the place for me to go.

You might well question the wisdom of the state of California in thinking of me as any kind of expert about America's largest bird. And I wouldn't disagree with you, to tell the truth. But nobody had made much progress in understanding the reasons for the decline of the condor. Just as with salmon, as I was to learn later, there were many theories, speculations, plausible stories, and folktales, but nothing anybody had tried had made any difference; the number of condors lessened. As the bird numbers continued to decline, there was a sense of desperation about the survival of the species. There was a desire for new ways of thinking, I suppose, and that's how I landed on this panel.

The five members of the "expert" panel were led by Jim Green, a Ph.D. in ornithology and a true scientific expert about raptors and vultures—eagles, hawks, turkey buzzards, and their relatives. The rest of us had considerable experience trying to puzzle out what happened to other endangered species and how we might make their situations better. Another panel member was Don Friedman, a mathematical-type, professional like Charley Russell—supercompetent and clear-thinking. Don wore a long black beard, had the kind of rimless glasses that John Bock-stoce (and John Lennon) sported, and looked as if he just stepped off a prestigious eastern university campus. Few who met him during our condor period knew that he had recently moved to Bozeman, Montana, to be on the faculty of Montana State University so he could get away from the crowded and expensive housing of Southern California and in-dulge in his favorite pastimes—hunting big game and hiking in the high-elevation forests of Montana. He did not look like a mountain man, nor the kind of wildlife expert who narrates major television nature shows, but he was, like Bockstoce, a tough outdoorsman and as such was some-one with much direct contact with nature.

At our first meeting, our leader, Jim, said that we had to go talk to Eban McMillan, a rancher without any formal education. Jim said that Eban knew as much about the California condor as anyone—he was *the* condor expert. Eban McMillan had no formal training, not in science, not in wildlife management, not in birdlife. He had spent his life as a dry-country rancher in Cholame, California, a town whose only claim to fame was a plaque announcing that it was there, at a crossroads in this dry country, that the movie actor James Dean had died in an automobile accident. Eban ranched a section—a square mile—in the golden hills of California inland from the ocean, within the rain shadow of the coastal mountains. His ranch was in semidesert terrain, with an average rainfall of about twelve inches a year.

So why did Jim insist we talk with Eban before anyone else? Eban and his brother, Ian, were considered to be the leading field experts about the condor, having spent whatever time they could seeking out the birds, watching them, and counting them. They were the longest-term and most careful observers of this bird and seemed to know as much about their abundance and changes in abundance as anyone. This sounded intriguing. The entire scientific panel appointed by the state drove up to Eban's ranch.

On our first visit to the ranch, I thought that it would be hard to find a place as different from the green dairy farms and forests of New Hampshire as Cholame. Eban took us around, showing us the brown grasses and scattered clumps of trees where cattle occasionally grazed. The dry air held the scent of desert shrubs: sagebrush, lemon bush, and ceonothus. Soft breezes sailed through heat waves that lifted a scent of a soil different from that of New England, a hard, drought-toughened, brick-forming clay. There was no hint of the perfume of mountain balsam fir, no snow-filled slopes.

"Don't run no more cattle than the land itself can carry. Don't irrigate; don't fertilize," Eban said as we walked in the heat of a cloudless fall day through his land. "Neighbors do just the opposite—fertilize, irrigate, run as many cattle as they can. Spend a lot more money. In a good year, they make more money than I do. But in a poor year, their cattle die and ruin the land—grasses suffer, soil erodes. My cattle carry through. In the long run—who knows—I do all right and this land that I grew up in still fits in with the rest of the countryside." So Eban believed in treating the land well and seeking what would later become fashionably referred to as "sustainability." It reminded me of my New Hampshire father-in-law Heman Chase's use of his woodlot to supply firewood to heat his home.

Eban was in his eighties: short, thin, wiry. A finger was missing from one hand, lost in some past farming accident. A native Californian, Eban reminded me uncannily of Heman Chase. They had the same body structure, had unusual—for the time—biblical names, and shared a common approach to and an appreciation for the land. They had similar lifestyles and philosophies: live close to the land; know it well; treat it well.

As Eban led us through his dry, rolling, tawny hills, we crossed several families of California quail, which I decided were one of the world's most charming birds—tiny, dainty ground birds, sporting a decorative topknot, a sort of circular decoration on a long stem, longer than the tiny bird's head. The quails scurried busily along the ground, the mother shooing the chicks along and looking out for them, the entire family making quick, excited dashes from one group of brush to another when they saw us coming, then pecking busily at the ground.

Here and there we crossed patches of dry land trees—California oaks in dry drainages—and groves of introduced Australian eucalyptus farther upland. We had to step over piles of dead limbs and branches in these woodland patches, sometimes climbing over large piles of brush, sometimes tripping and falling across them.

"Keep it scruffy—leave the brush for wildlife habitat," Eban said. "Takes a little land from grazing, but it's good for quail and other wildlife. Neighbors don't do it—clear out all the ground, try to farm every inch. Myself, I want to live with the land, want to see plants and animals." Yes, it was new country for me. And Eban did seem to be an expert about his lands and dry California.

As we walked and drove around the California countryside, I wondered what a meeting between Eban and Heman would be like. Would they instantly grasp how similar they were, or would it be a Californian looking at a New Hampshire man, with cultural differences forming a wall between them? Heman lived only seventy-five miles from the Atlantic Ocean, but didn't eat clam chowder. When offered some, he would say, "No, we don't eat that where I come from. They eat clams along the coast." He had been as far west as Wisconsin, but no farther, and was skeptical of anything as far away as California. I doubted that New Hampshire had ever been much in Eban's mind; his eyes were always on his own square mile and the California habitats of America's greatest bird. If the two met, would there be a meeting of minds, or would Heman be going in the kitchen door, Murray Buell style, while Eban was coming out the front door, California style? The main difference between the two seemed to be that Eban lacked the dry wit of New Hampshire, nor did I find that kind of humor elsewhere in California.

The dry California grasslands created a mysterious landscape, one in which you wouldn't be surprised to find a nymph running past you or some primitive god lurking in the hills, but I never heard anyone refer to it as a goddamned fairyland. Perhaps the desert countryside did not blossom that way, or perhaps there was some cultural tradition from Irish, Scottish, and English humor that disintegrated by the time Europeans settled California. Also, Heman was a lover of books, poetry, and writing, while Eban seemed to limit his pen to the condor. But both were experts about their own countryside.

Eban took us to locations where he used to see condors. He and his brother, Ian, together were said to have provided the longest continuous set of observations of this largest of all American birds. They were an epitome of a monitoring program, the two of them counting the condors every year, searching out why the birds were dying out, completely on their own without outside support, without research grants or governmental funding. Everyone we met who was involved in trying to save the condors knew about and respected the two brothers, especially Eban.

As we walked through his ranch, Eban talked about his experiences with condors. "Hard to explain why I've spent my life on condors," Eban said. "Well, tell you a story. Once I was down south on a big ranch in that rolling countryside overlooking the great central valley. Saw a condor soaring. Stopped and watched it come in. I lay down on the side of a grassy slope and watched the bird. Stayed still, and he got curious. Maybe I was a carcass he could eat. Took him a long time to descend. Finally hovered just over my head, maybe ten, maybe fourteen feet above my head. Could feel the wind made by his beating wings. Can't quite describe how it felt—touched by the breeze blown up by the biggest bird in North America. He hovered and I lay there, enjoying that wind. Powerful bird."

Eban stood silently, his mind soaring back to that magic moment. It wasn't an ideology that drove him. It was his love of the land and its wildlife, developed through years of living on it, with it, sharing himself with the soils, the plants, and the animals. It was clear to me that Eban was an incredible character, a rare resource—one of those people who watched and observed carefully, his observation filled with affection for the countryside but his mind open to the exactness of what he saw.

Why was the condor declining? That was what everyone interested in this bird wanted to know, including a large segment of the California public. What evidence there was suggested that the decline had been going on for a long time. Before European settlement, West Coast Indians had killed the bird for its long feathers. Lewis and Clark shot one, so they could describe its features, when they were on the Columbia River in 1805, but the condor had long since disappeared from Washington and Oregon and even Northern California. As long as Europeans had settled California and the Pacific Coast and anyone began to record counts of the birds, the numbers had steadily declined. About sixty condors soared over the sand-colored hills of Southern California when Eban and Ian began to count them.

If they were declining, but Indians were no longer shooting them, what was causing their decrease? There was a lot of controversy about that at the time our panel was formed. If the reason had been clear, nobody would have created a panel of scientists.

Eban was convinced that a widely used poison to rid the countryside of coyotes and rodents, called 1080, was one of the primary culprits. This antivermin poison was toxic to many animals, and it was persistent. Eban and others believed the poison outlived its victim and that if one of the great carrion-feeding birds ate the flesh of an animal that had, in turn, eaten 1080-laden bait, the bird too would be poisoned. There were informal stories traded around that someone somewhere had seen a condor feed on such a carcass and then watched it sicken and die. Eban told us several of these stories and repeated them at each of our meetings with him. But we could never pin down who saw this or where, nor could we find any evidence that the 1080 had actually been carried from a condor's prey and was found in its flesh. Nobody had made such measurements.

Others said that the birds were electrocuted when they perched on power lines. Such deaths had been observed for other raptors. Eban talked about how, when he sat at home on his ranch in the evening, once in a while the lights would flicker, and he would think that another condor or other large hawk or eagle had just shorted itself to death as its wings touched two wires. The beat of a condor's wings might cause its own death. A curious thought. But once again it was hard to find a demonstrated death of a condor from this cause.

Some said it was just hunters who didn't care and were amused to shoot such a big target. Some condors were shot, but that left the question as to whether this was a major or incidental cause of the general decline of the species. The word also went around that the birds were suffering from lead poisoning because they ate bullets, either inadvertently when they fed on some carrion, or—thinking that they were small pebbles—swallowed them with their crop. Nobody was sure.

The most commonly recognized experts on the condor were the members of the condor recovery team that had been formed by several environmental organizations. They were busy with their attempts to understand the causes of the decline of the condor and to figure out what they could do to help. Our panel met with that recovery team and found the members a mixed group. Some had no scientific training; one of the job experiences listed by a member of the team was his work as a professional clown.

The recovery team was involved in an elaborate set of actions that they believed would help save the condor; these actions also got them considerable publicity. One of their most expensive activities was their attempt to attach radio collars to each of the twenty-two birds remaining at large. The team had to find a radio transmitter small enough to mount on a condor, not an easy technological task at that time. They had difficulties with radios falling off condors, failing to transmit, and injuring the birds.

At one meeting with this team, we asked them why they were putting radio collars on the birds. They said that they wanted to track the birds to be able to determine the causes of death, so that they could respond to those causes and help save the species. I pointed out that there could be many causes of death and twenty-two birds were a small sample. By the time all the birds had died, it was unlikely that they would be able to characterize the major causes of death as any kind of average or dominant cause, and by the time half the remaining birds had died they would not have learned enough to do anything constructive for the remaining eleven condors. After considerable discussion, they agreed with this point and began to change their grounds for the major rationale for radio tracking, emphasizing knowing where the birds were so that they could begin to remove them from the wild and set up a captive breeding program. I believed I was right, but I don't think it made me especially popular with this better-known recovery team. As a matter of fact, whenever there was a discussion involving that group and ours, we won the intellectual discussion hands down, but with each verbal victory our standing with the condor recovery team declined. We were smart, generally well-educated, and able to argue and debate, but did that make us the experts?

So there were three sets of experts: us, appointed by the state; the recovery team; and the McMillan brothers. Our team concluded that whatever the specific causes of death of any individual, the condor habitat was in big trouble. "Better to have a small number of animals in a good habitat than a large number in a bad habitat," I said. Our panel focused on the habitat problem. Eban agreed with us.

The condor's habitat had been greatly altered since European settlement. Fires had been suppressed, and some speculated that the land that was once heavily grassland and was now shrubs no longer revealed carcasses to the condors, nor provided landing and takeoff strips long enough for this immense bird. Also, the wildlife that condors once depended on—deer and elk—were no longer abundant, so they were pretty much restricted to feeding on dead cattle. The birds were feeding mainly on dead calves and adult cattle on one very large ranch, the Tejon Ranch, that lay in the western part of California's great central valley. This ranch, so large that cattle who died on those slopes were left to rot, included eastward sloping, grass-covered slopes of the coastal mountains.

Part of the former range of the condor had included areas in the Los Padres National Forest, a large forest that extended to the hills behind Santa Barbara, a dry landscape with no more rainfall than at Eban McMillan's ranch—about twelve inches a year in some places, the same as a desert. Much of its landscape is now filled with tough chaparral shrublands, dense thickets with few places for the condors to land, poor grazing lands for wildlife, and few open areas for the birds to see carcasses. A large area in this national forest had been set aside as a condor reserve, with restrictions on its use. Unfortunately, the birds did not read the maps and did not use their reserve.

Another question that came up was: Why save this species anyway? Who cares? Our recommendations to the state were phrased in terms of what condors represented to people: Was there a necessary role for them in nature so that if the birds were removed, some other species or group of species would also disappear?

We were hardly the first to speculate about the role of the condor in nature—what ecologists today would call its "ecosystem function." In 1713, ninety years before Lewis and Clark shot a condor on the Columbia River, a British cleric, William Derham, published a book titled *Physico-Theology; or, A Demonstration of the Being and Attributes of God, from His Work of Creation.*[5] Derham was struggling to explain why, on an Earth made by a perfect God, vicious predators, such as the newly discovered Peruvian condor, existed. He called the condor that "most pernicious of birds" and "a fowl of that magnitude, strength and appetite, as to seize not only on the sheep and the lesser cattle, but even the larger beasts, yea the very children, too."

He noted that these scary creatures were among the rarest of animals, "being seldom seen, or only one, or a few in large countries; enough to keep up the species; but not to overcharge the world." Derham con- cluded that by a "very remarkable act of the Divine providence," the "useful creatures are produced in great plenty"—such as sheep and deer—while "creatures less useful, or by their voracity pernicious, have commonly fewer young, or do seldomer bring forth." And it was in this way that the rare predators were able to control the abundance of their prey, keeping nature in its balance. It was a statement of the great myth of the balance of nature, persistent in Western civilization for several thousand years, restated by Derham as a way of trying to understand the discovery of new animals through the exploration of the New World.

Derham's idea was appealing, but by the 1980s the condor did not seem essential to the persistence of other species or to its habitat. The demise of the condor had not led to eruptions of other animal species or other undesirable effects on the landscapes of Southern California. This argument had not gone away completely, however—people still argued that every species had some essential role and was irreplaceable in its ecosystem. Arguing that condors had some essential role in their ecosys- tem could be used to justify saving them, but that idea didn't seem to have much basis in fact.

You could also justify saving the condor because you thought its ge- netic characteristics might harbor something useful to us or to future ecosystems. In today's world of genetic engineering, one might say that perhaps some gene within the condor would be of great help to other species. Others suggested the condor was worth saving just because people wanted to see the bird, even if it were in a zoo. Still others argued that the condor had an innate right to exist, just as we humans do, and others said that the simple existence of this species, like all others, was satisfying to human beings and a sufficient justification in itself.

Our panel considered all these ideas. We met and talked with many people interested in the condors and asked them why they valued this bird. We concluded that the condor's primary value to people was its symbolism as a free and wild creature in a wild habitat—the great soaring bird beating a wind with its wings onto Eban McMillan.

So our slogan was: Save the habitat; let the condors fly free. We believed that captive breeding, little tested, would probably fail; and even if it did succeed, returning the birds to the wild would be a problem because a generation of birds would be born without experienced adults to guide them. If the condors were just in zoos, the wonderful idea that this great bird soared over some wild landscape would no longer be true, and our spirit and our imaginations would suffer. The idea that the bird would be confined forever to a patch in a zoo seemed not to mesh with the reasons people wanted to have the bird around. And if nothing was done to improve the habitat, there would be a future the same as the present for those birds reintroduced into the wild: nothing to eat and difficult landscapes to land in and nest in. We suggested that the other arguments for the conservation of the condor were minor and that available funds should focus on improving the condor's habitat, from a condor's point of view.

The state listened to our advice and then proceeded to ignore all of our recommendations, going with the suggestion of the condor recovery team. All the remaining birds were removed and the now famous captive breeding program was established. There would be no effort to improve habitat.

We had good ideas, we thought, but no one listened. So much for the impression we had made on the state about our expertise. And so much for the state's impression of Eban McMillan as well. As I said, the state

had three kinds of experts available to it: us with our university credentials; Eban McMillan with a lifetime of experience; and a recovery team
with not much of either.

Before the dust had settled, I had called friends in Washington, D.C.,
involved in the conservation of endangered species, and tried to get
them to talk with the Audubon Society and other groups, but they said,
"We're so glad somebody is doing something that we don't want to interfere." There was a pizzazz to the recovery team's ideas: birds in the zoo,
use of the most advanced technologies, testing new technologies to work
with this strange bird. Good stuff for the media. Our suggestions were
not so media appealing. We wanted the state to burn the chaparral and
turn it back into grasslands, to put out dead cattle carcasses, to run herds
of deer so that, as they died, the condors would find something to eat.
Not the stuff of a television show about nature.

A red-tailed hawk soared overhead and Eban and I talked once again
about what things we might have done for the condor. The light faded
and I was confronted once again with an experience in which I had met
a wonderful local character but failed to persuade a government agency
to follow what seemed to be the logical course of action.

Whenever I could afterward, I traveled to Cholame to visit with
Eban. Eban and I would sit on his back porch and talk about how to conserve the land and its creatures as we watched some California quail
scurry for cover.

Contrary to what we had expected, the captive breeding of condors
worked well, and years later a private organization called the Peregrine
Fund began to support the reintroduction of condors into the wild. In
the late 1990s, I went to see condors that had been released northeast
of the Grand Canyon near Marble Canyon, Arizona. We talked with a
young man working for the Peregrine Fund, who was sitting in a pickup
truck and talking by radio to a woman colleague up on the ridge. He
knew nothing about our attempts more than fifteen years before to try

to help the condor; I'm not sure he had ever heard of Eban McMillan. He spoke little and seemed rather uninterested in our presence—just another set of tourists interfering with his work, I supposed. We walked away from his truck and looked to the eastern sky. After about half an hour, we saw the huge wings of several condors sailing on updrafts off the distant mountains. To me, their wings looked about as long as those on a Cessna 152, a two-seater training airplane, but I suppose my mind exaggerated. It was an impressive and moving sight.

The condors had not yet learned to feed themselves and were supplied with food by the Peregrine Fund in the form of stillborn calves trucked in from Phoenix, Arizona. Not exactly the life of a wild creature in a wild habitat. These were driven up to a mesa summit where the birds roosted, up where the woman on the other end of the radiophone was watching the condors.

Some of the condors released in California had gotten into trouble and some had died, from exactly the kinds of problems we had envisioned they might. The habitat problems we had foreseen were confronting the birds now, and they were yet to be independent. Watching the condors soar, I hoped they would overcome these problems and learn how to live in their ancestors' habitats.

Here on the West Coast, on the opposite side of the continent from Woods Hole, I had run into the same general problem that had plagued me with sleeping whales and thirsty elephants: You can't always answer every question by yourself, so there are times when you have to rely on experts. In modern life, this seems to be the situation most of the time, surrounded as we are, immersed as we are, in high technology and the results of sciences we barely fathom. The problem is, how do you know somebody really is an expert? This time, the question was how the state of California might know and, more important, how our society in general might know so that we could actually do something to help these birds.

So who was the condor expert? In the end, from the perspective of today, it seems that the recovery team was right about some things: The captive breeding program worked. Our state panel was right about some things too. Attempts to reintroduce the bird into the wild were running into the problem we had forecast about their disappearing habitats. But as I watched the condors that had been introduced into Arizona, far from their original home, my thoughts turned to Eban McMillan, the old-time rancher who loved condors and devoted his spare time to finding the birds and counting them, year after year. His direct experiences with condors and his wisdom in approaching California's landscapes made him a special kind of expert about condors. The winds from the condor's wings gave him a spiritual contact that rose beyond our scientific expertise.

Nineteen

RACCOONS, PEOPLE, AND
THE GREAT CHAIN OF BEING

ne of the great ideas about nature in the classical world of the Greeks and Romans was the great chain of being. This is the idea that in the world there is a place and purpose for every creature, and in a properly functioning—truly "natural"—nature, each creature has a job to do; in nature undisturbed, each does so, diligently, reliably. Of course, in the ideas of Western civilization, human beings occupy the most important position in the chain. Raccoons, I have discovered, have their own ranking of the relative role and importance of creatures. Although it is different from that of the Greeks and the Romans, ranking probably is just as important to them as it was to the classical philosophers.

When I was teaching at the Yale School of Forestry and Environmental Studies, I lived with my family in an old barn converted to a rustic house in the small suburb of Woodbridge, Connecticut, about nine miles west of the Yale campus in New Haven. The barn had a dug-out,

dirt-floor half basement, an unmortared stone foundation, and two stories of rough, wood-paneled pine walls and wood-floored rooms. Behind
the old barn, second-growth forests extended several miles westward on
rugged hills of granite, gneiss, and schist, where heavily eroded garnets
could be pulled from the outcropping rocks. In the front, a small stream
bubbled past the house-barn and a small stone dam built by some former
owner formed a shallow pond where skunk cabbage and touch-me-nots
grew and frogs croaked every spring. Wildlife was abundant, and it was
a wonderful place to live with two small children.

Life in the converted barn wasn't always perfect, however. One of
the banes of our lives was raccoons, especially when they decided to get
into our garbage cans, eat whatever food appealed to them, and scatter
trash around the backyard. This was a common problem. Various devices could be bought at the local hardware store that were advertised
as ways to raccoon-proof garbage-can lids—big spring-loaded latches and
special designs of the garbage cans themselves. None of them thwarted
our raccoons, who always found a way in. As annoying as the scattered
garbage and trash were, the loud noises the raccoons made as they pried
the lids off the cans were even more bothersome, especially when I was
trying to get some sleep.

One winter evening not long after dark, a family of raccoons attacked
our garbage cans. I took a large flashlight—the kind powered by a lantern
battery about four inches square and six inches high. Standing at an
upstairs window, I located the raccoons and fixed the powerful light
on them, hoping this would drive them away. The raccoons backed off
about six feet and looked at the light for a few minutes. When nothing
else happened, they returned to the garbage cans and began banging
them. I realized that the raccoons could see nothing threatening when
the light shone directly on them because the light blinded them.

So I turned the flashlight on myself, hoping that the clear image of a
person would scare them away. The raccoons stopped and backed off

about twelve feet. They watched me for a few minutes, then returned to the garbage cans, ignoring me. As they did, I noticed our cat sitting on a woodpile about a half-dozen yards from the garbage cans. He was a large cat, weighing about sixteen pounds, and his fur was gray, so he did not show up well at night. I turned the flashlight on the cat and the raccoons immediately ran off into the woods and did not come back that night. Their actions had made clear how they ranked my relative importance. In the raccoon's great chain of being, cats sat in a higher and more important position than a mere person, who was easily outwitted and was rather slow and unagile compared to the raccoon and the cat. Raccoons were experts about threats; to them, the cat was the expert threat. Not only wasn't I able to solve human-selected environmental problems, I didn't seem to rank very high within the animal kingdom either. It was a sobering thought.

Twenty

≈

HOW THE FOX
CAUGHT THE SQUIRREL

n a June evening in early 1973, I set out in a canoe on a
long inlet of Washington Harbor, at the western end of
America's Isle Royale National Park. One of the lesser-
known and least visited of the national parks of the United
States, Isle Royale is a remote 280-square-mile patch of forests, streams,
bogs, and interior lakes in the midst of Lake Superior near the border between the United States and Canada. I had done research at this park for
several years and would continue to do so with my colleague at Yale, Peter Jordan, an expert on moose. We were trying to understand how the
moose fit into their surroundings. It was research in one of the least disturbed wilderness areas in the United States, if not in the whole world.

Soon after I had pushed out from shore, a large bull moose strode
from the cedars and firs and stepped carefully into the cold lake waters,
where he began a slow traverse of the shallows, his head under the
wavelets as he searched for water irises, lilies, and other water plants
that were some of the favorite foods of this large northern ungulate.

The end of the inlet, just east of the moose and myself, was elegantly marked by an old cedar arching gracefully over the waters, framing the forest and the deepening sky beyond. The serenity and beauty of the scene rivaled the best of America's landscape paintings, reminding me of Albert Bierstadt's famous painting of the sunset at Yosemite and the nineteenth-century Hudson River School of forest scenes without human inhabitants. For that moment, the remote island wilderness appeared as tranquil as a still life, and as permanent in form and structure as brush strokes on canvas at the National Gallery in Washington, D.C.

Isle Royale is one of the best examples of true wilderness—of nature undisturbed by human influence—remaining anywhere in the world, but its real character is far different from the still life we imagine nature to be. The greatest of the islands in that greatest of all lakes, Isle Royale was named by French *voyageurs* who passed by her shores, visiting her only occasionally and briefly on their way to the north woods to trade for furs with the American Indians. Few ever lived on the island, and none disturbed her forests and streams in any lasting way.

It was this lack of direct human influence that had attracted me to Isle Royale. As an ecologist, I was interested in the workings of "natural ecosystems," a twentieth-century word for "nature." I was trying to understand what natural forces might determine the abundance of the moose, its vegetative food sources, and its only natural predator, the North American timber wolf.

At the time that the moose and I shared Washington Harbor, the standard belief among my colleagues was that a pure wilderness like Isle Royale would achieve a permanent, constant condition without human influence, and if temporarily disturbed by fire or storm, would recover its original status. In Isle Royale's case, the natural constancy was supposed to include exactly the right number of wolves feeding on just enough moose to keep both their numbers constant, or perhaps oscillating regularly like pendulums in two grandfather clocks. The moose would in turn eat just the amount of vegetation that grew in any year, so wolves, moose, and vegetation would continue to exist in a constant abundance, providing neccessities for each other and the controls needed to maintain mutual constancy. I had come to the island to discover how this ideal condition might come about. When we weren't doing our formal work, we were out nature-watching, which happened just about everywhere, all the time. Twilight was a good time to do a little nature-watching.

The moose circled the shallows for perhaps twenty minutes, putting his head down into the water and pulling it back out, water dripping from its bearded jaw. He rarely stopped to feed. In this northern wilderness, June was much too early for water plants, and as the moose edged his way over to the north shore he clearly found little on the bottom to satisfy his appetite.

Suddenly, the bull moose galloped through the shallows, scrambled out of the inlet, and began kicking vigorously at the shore. He dashed up a short bluff, breathing rapidly, turned, raced down, and kicked again where the sand and waters met. It was as if he were furious with the harbor for denying him food and was taking out his frustrations at the strand. What was the moose actually doing? I wondered. Moose seemed to be among nature's odder creatures, somewhat unpredictable, somewhat strange. Later Peter told me that, several years before, he had been hiking through the island's forest and came across a moose in a small clearing. He stopped and watched. The moose began to gambol about, jumping into the air, as if it were dancing. But there were no other moose in view. It didn't make any sense to Peter.

Another time, we were camped along the shore of Washington Harbor when a big storm came up, blowing so hard that cedar trees along the shore bent almost horizontal and waves more than three feet high crashed against the shore. Peter and I were out walking in the storm, enjoying this bit of wild weather, when a bull moose came out of the for-

est, looked briefly straight ahead, and plunged into the waves. He swam
a quarter of a mile to the other side; it was rough going for him even
though moose are good swimmers. He made it, but it seemed to me that
he had plenty to eat on our side of the harbor and could have mellowed
out for a day until the storm passed. Why he chose to swim in the
wildest weather I had ever seen on the island I have never fathomed.

The moose kicking at the shore is an image that has stayed with me
since. Nothing could have contrasted more with the idyllic scenery of
that evening than the moose's bizarre, chaotic, and perplexing behavior.
But in the quarter century since I watched that bull moose, I have studied nature's character, and have come to realize that the seeming constancy and stillness of Washington Harbor and its surrounding forests
symbolized a false myth about nature. In contrast, the moose that kicked
at the shore—complex, changeable, hard to explain, but intriguing and
appealing in its individuality—was closer to the true character of biological nature, with its complex interplays of life and physical environment.

Surprisingly, moose are relative newcomers to Isle Royale, first ap-
pearing there at the beginning of the twentieth century. It's surprising
because the island seems to be a natural, even ideal, habitat for them.

According to classic scientific theory, first suggested in 1839 by the
Belgian scientist Pierre-François Verhulst, after its arrival the moose
population should have grown continuously and smoothly, following an
S-shaped pattern known as the "logistic growth curve," which eventually brings the population to a constant maximum abundance, called the
"carrying capacity." But the moose population did not follow that nice
theoretical curve. It expanded rapidly to an estimated three thousand or
so, and then, overconsuming its food supply, crashed in the 1930s to perhaps as few as four hundred.

Wolves are even more recent immigrants to the island, arriving in the
late 1940s. According to classic ecological theory, after their arrival the
wolves should have held the moose population in check so that the two
species oscillated regularly, with the moose reaching its peak at the time

the wolves reached their minimum, and vice versa. But the populations did not follow those theoretical patterns, either. The moose and wolf numbers varied, but irregularly, not according to any simple rule that we could discern.

Since the evening when I canoed in Washington Harbor, I have searched the scientific literature for examples of constancy in nature but have found only the contrary: histories of populations that varied greatly over time. The longest records for estimates of the abundance of mammals are those kept by the Hudson's Bay Company, whose fur trading in northern Canada began in the early seventeenth century. The number of fur pelts purchased by the company varied over a tremendous range from year to year, decade to decade, century to century. Studies of these records suggest that they reflect variations in the abundances of fur-bearing animals—lynx, hare, and so forth—rather than economic conditions. All other cases in which the numbers of animals have been monitored over time show widely fluctuating change.

At the same time, during much of the twentieth century, ecology theories about animal populations assumed and predicted constancy. Botanists who studied forests believed in a similar fashion that a community of trees would grow from a clearing to a maximum abundance, called an "ecological climax," that would persist indefinitely without disturbance and, if disturbed, would return to that same constancy.

This idea of constancy was taken literally. In the first decades of the twentieth century, one American ecologist, William Cooper, carried out a classic ecological study of the forests of Isle Royale and concluded that they were in the kind of equilibrium I have just described, a conclusion he reached, curiously, "in spite of appearances to the contrary"—that is, in spite of obvious signs of recent change.

According to this classic expert, nature was supposed to be supremely ordered and logical, functioning as well as the best water-powered mill in New England, all the gears—or species—playing their essential roles. But that kind of expert's idea, so dominant in our civilization, did not seem to fit the bizarre behavior of the moose who kicked at the shore. As I watched him, I imagined that moose was an expert at something, but I certainly had no idea what. He seemed to epitomize an important aspect of nature that I had had difficulty explaining—either to myself or to other people—that inherent in nature was at least a little that was confusing, complex, and, well, downright strange.

I was thinking about that moose on another warm summer afternoon at Isle Royale National Park. We were taking our usual Sunday off from our work. I was relaxing at our research campsite at the west end of the island. It was surrounded by trees but had enough open ground for a set of camping tents and a tented mosquito-netting "living room," within which was our dining table, a crude picnic table. It was a short walk from our campsite to the harbor.

On this Sunday, everybody else had gone away, and I was enjoying time alone. I neatened up my tent, if you call the status of the office I would later occupy at Woods Hole neat, a fairly unusual activity, and then sat outside and began writing letters. I noticed a red fox sitting near a trail not far from my tent. Foxes are among my favorite small mammals. They not only are beautiful but seem smart and sometimes appear to have a sense of humor. Our recent experiences with foxes had come about because of their love of anything leather. A family of foxes had invaded our campsite a few nights before and grabbed somebody's wallet and several other small leather items, which they fought over, making so much noise that we all woke up. Seeing what was happening, we chased them and managed to retrieve the wallet, a little the worse for having been chewed on, but minus only a few items.

This Sunday morning, the solitary red fox quietly watched two squirrels in the trees. At first, the squirrels saw the fox, became wary and stopped their chattering. Motionless, they eyed the fox. But the fox did not move. After a few minutes, the squirrels forgot about the fox and went back to their chatter. Red squirrels are territorial, and one was trying to move into the territory of the other. The owner of the territory began to chase the intruder. They went round and round, in the trees and on the ground. The fox sat silently and watched. Only his eyes moved. He seemed alert but relaxed.

I stopped what I was doing to watch the drama unfolding. The squirrels' movements took them nearer and nearer to the fox. Their chatter became more intense and their battle over territory more and more vigorous. Squirrels are not nearly as smart as foxes and seem to have a shorter attention span. The issue at hand for them was no longer the fox, but who would rule the squirrel territory. The fox watched and watched, using no energy, wasting no effort. He yawned and stretched leisurely in the sunlight. But his eyes never left the squirrels, nor did he move in any other way. Eventually, the home-based squirrel chased the intruder right past the fox, within a few inches of him. Still they did not notice him. As they passed, the fox reached out a paw and, with a smooth, quick, and quiet motion, grabbed the second squirrel and ambled off into the woods with his lunch.

The fox was a model of efficient predation. He was also a model of good nature-watching. Watch quietly and do not let yourself be distracted. This is the way to learn about nature. The fox knew this well. He was an expert observer of nature. As with this fox, sometimes the best way to get what we want from nature is to observe and wait. In this case, lunch came to the fox that watched well and waited quietly. In contrast, the moose that kicked at the shore, whatever his kind of expertise, did not seem to get what he desired. Perhaps expert abilities differed among the animals of the world, just as they differed among people. I didn't know. I sat listening to the sound of loons calling over the water, a haunting, crazy sound that seemed to fit the occasion.

Twenty-one

ROLL ON, COLUMBIA:
IDEAS AND BUREAUCRACIES

he room in Gold Beach, Oregon, was crowded with commercial salmon fishermen and fishing guides, all looking at me with hostile eyes, their arms crossed, their backs upright or their bodies leaning forward aggressively. I had just come in from a twilight walk along the mouth of the Rogue River, one of Oregon's most famous salmon fishing streams. The wind had blown powerfully, tossing spray from whitecaps, salting my face. Sea lions and harbor seals barked in the waves, only their heads visible, as if they were sitting on spring-loaded chairs that raised them up and down with the waves. Their demeanor seemed friendly compared to the fishermen's faces in the crowded, overheated room.

At Gold Beach, the Rogue River splashed its way to the ocean, mixing roughly with the Pacific, neither body of water seeming to give way. I had walked along a breakwater and on the rocky shore. I had looked down into the Rogue River hoping to see salmon, but in the almost horizontal sunlight breaking through storm clouds, I could not see below the sur-face. I stood in the wind, hoping it would cleanse my mind as well as my skin, to help me get ready for the public meeting I was about to direct.

What was I doing here? I wondered. It was the early 1990s and I was on the far reaches of the Pacific Coast, in charge of a study of salmon, a species I had little knowledge about until the state of Oregon had ap-proached me about running a project to study the relative effects of forestry on the decline of salmon, or the perceived decline of salmon. As part of my latest attempt to see if I could solve any environmental prob-lems, I had decided that this project would include open public hearings, and the one inside the building I was about to enter was perhaps the most crucial.

The state agencies had warned me that public forums would be use-less. "All that will happen is that the environmentalists will yell at the antienvironmentalists, and vice versa," one of them had said. "Waste of your time."

And acrimony seemed to have started already. Gold Beach was one of the centers for sport salmon fishing, and its constituents were crucial to our work; we needed to have them on our side. But a few days before, the head of the salmon fishing guides association of Oregon had called my office and complained.

"You've set up a public meeting during the day," the person said, "but we're all working people. Can't take time off during the day to go to some meeting. We want you to change the meeting time to an evening."

Susan Day, the administrator of the project, called me. "What should we do?" she asked.

"What the heck," I said. "Let's have two meetings. One the night before for the fishing guides, another the next day as planned and advertised."

Now the moment of reckoning was at hand. The five scientists I had persuaded to join me on an expert panel were sitting inside. They were my friends as well as my colleagues, and I felt responsible for them. If the meeting was hostile and a disaster, I would feel bad for them as well as for myself. A hostile meeting with the fishing guides could ruin the entire project—if the guides got mad enough at us to seek out their legis- lators and complain. The wild Pacific seemed, at the moment, a calmer place than the inside of that building. Well, the time had come, so I turned away from the onshore wind and walked into the crowd. It was hot, almost steamy. I took my place at the head of the table. It struck me that my experience in confronting wildlife was going to be helpful. Be calm; don't show fear; be careful whom you looked in the eye and whom you didn't. I took a deep breath of the sweaty indoor air and began the meeting.

"Professor Botkin." A voice broke in before I had gotten more than a sentence or two out. "I'm the head of the Oregon Fishing Guides Asso- ciation. Do you believe that salmon in Oregon are actually declining?"

It was an important question, but it also seemed to be an aggressive assertion by an alpha male showing his power in front of what he saw was another alpha male. Not the best metaphor, but one that came to mind at the moment. I looked around the room at the hostile people. I looked the speaker in the eye. "We've just started this project," I said. "We're just here to learn. We haven't made any assumptions about anything."

Susan Day, watching from behind me, said that there was a collective exhale from the entire crowd and the postures around the room shifted. Arms unfolded, bodies relaxed, and people leaned back in their chairs. Expressions became open and friendly. A friendly conversation ensued.

A little later, a thin, wiry elderly man stood up, a black patch over one eye. "Name's Lee Hill. Been a fisherman my whole life. Just want to say that I have no scientific training, but it just makes sense—if these salmon are spawned and reared in a freshwater stream and then go to the ocean for three years and return—it just makes sense that the water flow in the year they were born should have a big effect on how many salmon come back." He paused while a friend brought out some props.

"As I say, I'm not a scientist. But I went to the U.S. Geological Survey and got their water flow records for the Rogue River right out there"—he pointed to the outside—"and the Umpqua River. Then I went to O-D-F and W (the state's fish and wildlife agency) and got their counts of salmon on these rivers. Here's what it looks like."

His friend brought out a huge board with the two factors graphed, water flow and counts of adult salmon swimming upstream each year, with the year on the bottom of the graph. Scientist or not, old Lee Hill had looked at quantitative data. The audience looked at the graphs. The graphs were impressive. A high-water year was followed three years later by a high salmon return, a low-water year by few salmon returning.

A lot of discussion followed, relaxed and animated. At the end of the meeting, the head of the fishing guides spoke up again. "Look," he said, "we fishing guides are on the rivers three hundred sixty days a year. We know the rivers better than anyone. We're prepared, Dr. Botkin, to offer to help you. We're willing to make any measurements that would be helpful to your study." A complete turnaround. The public forum had become incredibly successful; in addition, we had a great insight from eighty-year-old Lee Hill that showed a strong correlation between an environmental factor and salmon.

Lee Hill had no ranking, as our society views these things, as a fisheries scientist. But throughout this salmon study, we continually spoke with fisheries scientists, hydrological scientists and engineers, and government agency personnel. Another part of our project dealt with what would have seemed to be the opposite end of the expertise chain—the engineers and scientists at Bonneville Power Administration. One of the ways we became involved with BPA people was over their methods to forecast possible fates of salmon.

I had grown up with the idea that the BPA was a federal agency with a socially beneficial goal, an agency that I would expect to find useful and sympathetic. Part of the reason for this was Woody Guthrie.

Born July 14, 1912, in Okemah, Oklahoma, Woody Guthrie became one of America's greatest folksingers and folksong writers. He wrote "This Land Is Your Land" and hundreds of other songs that are part of America's campfire sing-alongs. He was an iconoclastic do-gooder, a creative genius, a labor organizer, a political radical, a hobo and bum, of whom one of his cosingers said of the young Guthrie, "You wouldn't want to run your hand through Woody's hair, because you'd never know what might crawl out of it." Years after his death, he remains a folk legend and a genuine American character. Woody survived the Great Dust Bowl era in Oklahoma, home fires that killed his kin, and knew the Great Depression from the viewpoint of hitchhiking on freight trains and bumming around the country, from listening and talking to the common folk. He believed, like Carl Sandburg, in "The People, Yes." There is no doubt that Woody Guthrie wanted to do what was right and believed that he *knew* what was right.

I met Woody Guthrie once. When I was eight years old, he came to our house to visit. He carried a banjo and played some tunes. He was a short, wiry man, with an open and friendly manner with children. Woody and I sat outside under the shade of a tree and played mumblety-peg. Others found Woody tough—so idealistic as to be annoying. But he wasn't like that on that day with me. He was just a simple grown man playing an old game with a small boy.

During the Great Depression, President Franklin Roosevelt set up the Bonneville Power Administration to build dams on the Columbia River system, including the Snake River. It was part of the public works projects that the Roosevelt administration put together to provide jobs for people out of work. It was also part of that administration's attempt to develop the nation's resources to the benefit of the people. At the time, it seemed a very good idea.

In the 1930s, when Woody was already famous but still quite young, he was hired by the nascent BPA—it is said as the agency's first employee—to write folksongs about the new Columbia River dams and the Columbia River projects. Woody was hired to do what we call today public relations, to get the word out about the Bonneville projects in a way that appealed to the people and made them acceptable to the nation. Woody wrote twenty-two songs for the BPA, all of them celebrating the building of the dams and what they would do for the people. One of the best known of these songs is "Roll On, Columbia," whose chorus contains the line *Your power is turning our darkness to dawn.*

The first of the great dams BPA built on the Columbia River system was the Bonneville Power Dam. Woody's song celebrated that dam. In one of the stanzas, Woody wrote:

The waters have risen and cleared all the rocks,
Shiploads of plenty will steam past the docks,
So roll on, Columbia, roll on.

The biggest of the dams, and the farthest upstream on the Columbia River, was the Grand Coulee. About that dam, Woody wrote:

And on up the river is Grand Coulee Dam,
The mightiest thing ever built by a man,
To run the great factories and water the land,
It's roll on, Columbia, roll on.

In "Talking Columbia Blues," Woody sang, "*I don't believe in dictators, but one day this country's gonna be run by e-lec-tricity.*" He praised the idea that in the future everything would be made of plastic and that the electric power from the dams would provide light for the milkmaid and the aluminum industry. In another of his songs, he wrote that King Salmon was going to have to give way to the big dams. It's too bad, but that's the way it has to be, Woody's lyrics said. The big Douglas fir trees would have to be cut as well. That, too, was too bad, but it had to be, his lyrics said. When Woody Guthrie wrote songs for the BPA about the benefits of the big dams, he believed that they were good because they would provide jobs for people.

Among folksingers of that era, a well-known story about Woody is that he was offered a high-paying job playing folksongs on a radio show. He accepted until the producers told him that he would have to wear cowboy boots, jeans, and a hat. I'm not a cowboy and I won't dress up like one, he told them. When they insisted, he quit. Woody Guthrie was not a person to sell out his ideals. In the 1940s, he joined with Pete Seeger and Woody Hayes to form the Almanac Singers, a group that went around using the power of their music to help organize unions.

The involvement of Woody Guthrie with the BPA suggests that people who believed in the people, especially in the common folk, also believed in the 1930s that the BPA was a good idea, not just to big government, not just to the big power and aluminum industries, not just to the ranchers, but to the people everywhere. It was seen as a progressive and innovative idea.

But by the 1990s, when I was standing at the mouth of the Rogue River in Oregon watching sea lions and harbor seals, the BPA had become a large quasi-government bureaucracy providing electric power and irrigation water. Its large dams had had immense effects on the environment. Thousands of miles of streams had been altered.

We were told that the BPA had created the most sophisticated computer simulations dealing with threats to salmon. One of the members of the panel I organized to study the salmon was a long-term colleague and friend, Matt Sobel, an expert in applied mathematics and in operations research. At the time, he was dean of the School of Business at the State University of New York in Stony Brook, Long Island. Matt and I ana-

lyzed these supposedly sophisticated computer models. One was supposed to examine the effects of deaths of salmon when they passed
through the dams, occurring as a result of injuries they suffered from the
dams. I called the BPA's computer model the pinball machine salmon
game. In this computer simulation, imaginary young salmon swam downstream and there was a chance that they would bang into one of the BPA
dams. If they did, they were dead—out of the game, just like the little
metal balls in a pinball game. There was nothing else to the game that
could affect salmon. Everything else about the environment was assumed to be constant. The scientists who created this computer game
had bought into traditional ideas about fisheries management, ideas that
went back seventy years but had long been proved wrong. The underlying idea was the belief in the balance of nature that I mentioned before.
It affected how people viewed predators and was the basis for the idea
that nature, left alone, achieved an unchanging state—that the abundances of all creatures would remain the same indefinitely. And if the
abundances were temporarily disturbed, they would recover back to
their exact natural abundance—nature's carrying capacity. Fisheries scientists had created mathematical models that assumed that this was true.
The pinball game for salmon accepted all these assumptions.

It was a strange way to think for a bureaucracy trying to protect itself from blame. Suppose you were a lawyer for the BPA trying to make
it look as if the agency's effect on salmon was unimportant. To do this,
you would hire scientists to create a model with all the environmental
qualities that they could think of varying—water flow, temperature,
amount of gravel, ocean currents, numbers of predators. With all these
things changing all the time, the effect of a few young salmon banging
their heads against the dams would appear minor. Instead, the BPA scientists had created an imaginary model that made the BPA appear to
have the greatest possible effect on the salmon. The irony was that the
assumptions were wrong. But the sad consequence was that, locked into

it had not worked out quite the way it was supposed to. There were unintended consequences, including the risk to the salmon and the extent to which people would care about it. A large bureaucracy seemed limited in its ability to deal with nature, not able to sponsor the kind of creative insight that might come to a longtime fisherman who had a natural sense of curiosity and a bent toward quantitative analysis. Lee Hill was like Eban McMillan and Heman Chase, all three careful observers of their surroundings, careful thinkers about it, fascinated by it, and trying to do their best, working mainly on their own or in small groups. Maybe it was better to be like the fox, acting alone, than like the squirrels at Isle Royale, confused by their internal squabbling and concerns.

Twenty-two

≈

PEANUT BUTTER IN SPACE

 ack one day in the 1970s, long before I even thought I would be involved with condors and salmon, I was trying to clean up the piles of papers scattered around my office at Woods Hole. In the middle of a moderately deep pile of unread mail, I came across an article that had gathered dust from the sandy cape and salt spray from the ocean. I had glanced at it briefly once before. Now it caught my attention. It was about a plan to build a giant space station orbiting the Earth, one that would hold ten thousand people who would live, according to the article, in a kind of Garden of Eden. The space station was to be shaped like a giant doughnut, a shape known to mathematicians as a torus; the illustration in the article showed a cutaway view of life in the doughnut, making the huge construction appear as a horn of plenty. Inside, people floated through the air. There was an equivalent of gravity created by the slow rotation of the torus, but the force would be slight enough so that you could fly. People flew and landed on plants, like small birds, feeding on fruits.

My office was in sharp contrast to that soaring Garden of Eden. It was a long rectangle with little character and no charm. It had black soapstone counters running the entire length of one interior wall and, at right angles, along the shorter wall whose windows looked out at Woods Hole. The room had been built as a small laboratory, with a sink in one corner and plenty of room to spread piles of papers and books from one end of the L-shaped counter to the next. The walls were the color of public restroom walls, dull and uninviting.

For a moment I glanced at the unfiled papers, and then, thinking better of trying to neaten up. I settled into my office chair and read the article. The author, a professor of engineering at Princeton University named Jerry O'Neill, wrote that the biological problems of building the huge live-in space station were simple. All you had to do, he wrote, was sterilize everything and get it up there.

Space travel had been and still is one of my longest fascinations. When I was five years old, my sister, Dorothy, and I had cut out circles of different sizes to represent the sun and the nine planets of our solar system. We drew and cut out paper spaceships and flew them from planet to planet. The necessity required to keep life going in a spaceship seemed clear to me then—water, oxygen, food, a place for waste, a way to recycle. I spent hours designing the best places to store oxygen, water, food, and have room for people on a spaceship.

Reading Jerry O'Neill's article amid my Woods Hole clutter, I thought that he had missed some major points. We ecologists down on the Earth weren't even about to help save the whales or even figure out their social behavior. We didn't know how many hours a whale slept or how much water an elephant drank. Try as we might, we just didn't seem to be very successful in solving environmental problems here on our home planet. And yet here was an engineer saying it would all be simple up in space. I didn't believe it. How was he going to make sure that the oxygen supplied by growing green plants was enough for the activities of all those people floating and eating fruit off the trees? Those people who required oxygen and breathed out carbon dioxide? Would the plants take up the carbon dioxide fast enough to avoid that gas reaching toxic levels in the space station's atmosphere? This was just the first of many problems.

I pushed away the papers around my telephone and called Lynn Margulis, a biologist famous for her work on early life and for her contributions to the idea that became known as the Gaia Hypothesis, which she worked out with James Lovelock: the idea that life had influenced the entire Earth's life-support system—all of the atmosphere, the oceans, and much of the solid crust and soils of our planet. She had planned a visit to Woods Hole anyway, I discovered, and we agreed to meet to discuss Jerry O'Neill's article when she came. She called Rusty Schweicker, then an active astronaut, and suggested he join us, which he did. He flew up in an astronaut plane, able to do so because he was required to maintain a certain number of flying hours and there was a military airbase not far from Woods Hole.

The three of us met in an office at the Marine Biological Laboratory, Rusty in his flight suit with an emergency knife in a sheath attached to his right leg. We went through the article and decided that we should not let this pass. At the time, Jerry O'Neill was getting quite a bit of publicity about his idea. There was a special television program that showed him flying away in his private airplane, discussing space stations. And some people were suggesting that the solution to the human population problem was to place all the excess people in O'Neill-like space stations.

We wrote a brief article discussing the challenges that confronted the life-support system for such a floating doughnut filled with people. Lynn took it on a trip to Europe and returned with the signatures of about twenty other scientists. We got it published in an obscure scientific magazine and one popular magazine.

By the time all this had happened, the piles of paper in my office had built up to a danger point and I had decided it was really neces-sary to clean up. While I was trying to neaten up, the phone rang. The caller said that he was chairman of the National Academy of Sciences Space Science Board, and that the board was holding a summer study to discuss life sciences. He explained that he had read the article we had written about space stations. Would I be willing to conduct a sum-mer study about the ecological problems associated with long-term space travel?

With my lifelong fascination with space travel, it was an offer I had a hard time turning down. I thought about it for a while. There were two kinds of studies that could be done. One would be a straightforward en-gineering study: How many tons of water, oxygen, and so forth would it be necessary to carry aloft? That did not seem very interesting; it was a kind of accounting. The other kind of study was to ask if it were really possible to create an artificial ecosystem that would sustain itself. This would take a lot of imagination, and it might be another way to get at the kind of understanding of nature that I had tried so hard to attain. If you could create an ecosystem from life forms of the Earth, then you would have to understand a lot about nature. The process of developing the space station would increase our understanding of ecology.

I called the head of the Space Science Board and explained the two possibilities, saying I would be willing to do the second. He agreed. I de-cided that the path to this study was to find the most imaginative people I could—and the broadest thinking. I asked Harold Morowitz, a bio-physicist who had written about ecology and was one of the smartest and most widely read people I knew. Harold understood biochemistry as well as biophysics, knew a lot of mathematics, and had written about the limits that energy placed on life. He agreed to join the study.

Then I called Larry Slobodkin, an ecologist famous not only for his work but for his breadth of knowledge of many subjects and his incredible imagination. Then I found a mathematician, Barry Little, who turned up at my door at Woods Hole and told me he had decided to shift his work to ecology and asked if I knew of any projects that might interest him. I immediately invited him to participate.

The work needed a microbiologist, because bacteria and other tiny creatures could have huge effects on any ecosystem, but especially on one floating in space. I knew from Lynn Margulis that some bacteria play crucial ecological roles, fixing nitrogen from the atmosphere and returning organic nitrogen to the atmosphere, and doing many other essential chemical tasks. Other bacteria could cause an epidemic either among the people or among other animals or plants. Lynn suggested a colleague of hers, and he agreed to come. Larry suggested an ecologist who had created and studied tiny closed ecosystems, and he joined the team.

We five assembled on a beautiful summer morning in Snowmass, Colorado, where the Space Science Board was meeting. We met in a ski lodge room and took brief walks as breaks from our discussion. Sublime mountains rose in the distance, where forests formed deep green streaks as if their colors had slid down the slopes. We felt soft winds from the mountains and smelled sweet scents of conifers. Squirrels chattered and birds called. Snowmass abounded with life; the air was a pleasure to breathe; it was about as different from an enclosed doughnut-shaped orbiting space station as I could imagine.

One day everyone took the afternoon off, and I took my family to Maroon Bells, a famous Colorado high mountain valley, where we followed a hiking trail over a low pass to a hidden lake. On the way, we passed a mountain meadow where yellow and red flowers bloomed in the brief summer, and tundra grasses stretched into the distant forests. Landslides on the steep mountain slopes brushed away streaks of forests, painting the land gray and tan. How could one improve on this? The doughnut in space with ten thousand residents living in a man-made

Garden of Eden seemed an impossible and arrogant idea. What would the scents and sounds be like in that torus with ten thousand people spinning in space above the Earth? Would they be as sweet as the scent of fir, as bright as the blossoms of Maroon Bells?

Back in the ski lodge, we five sat down and began to talk. Larry, a tall and heavyset man with a long beard, squeezed himself into one of the chairs in the small room. He said, "Imagine this, this space station. It's going to have a control center full of knobs and dials. There will be people manning that control room. They will read the dials and adjust the knobs. Now, the first thing they will want to have is a dial that tells them that the whole system is about to fail."

"You mean sort of like the oil pressure gauge on a car," I said.

"Exactly," said Larry, "or when the doughnut will get a flat tire."

We began to talk about what this dial might be. What was the equivalent of an oil pressure gauge for a huge, closed ecosystem that was supposed to support people?

Harold Morowitz began to draw diagrams on pads of paper. He drew boxes and arrows representing the flow and storage of chemical elements needed by living things and how these might move around the space station. We talked throughout the day about Larry's gauge without getting any kind of answer.

"You see, if it were an automobile that we had designed ourselves from start to finish, we would know what variables were crucial," said Harold. "But we didn't invent life and we don't understand ecosystems, so we don't really know what that gauge would be like."

We quickly decided that the only way to approach the question of what the dial would be was to think about how to create a scientific theory for ecological systems. We decided to take a biological approach and suggested that the names of the species in the system could serve as a foundation for what the system was like. If there was an elephant in it, it would be dry land and there would be no whales. Each species named told us a great many species that could not be there. A beginning list of who lived in it would constrain what the system would be like. Knowing this, the list of possible dials became much shorter and we thought we might find a dial, or perhaps a set of dials, that would work. This was the beginning of a long series of conversations that continued over five years.

We thought this was interesting and might even be important, but our colleagues thought we were crazy. We gave presentations about our ideas and were able to publish them in obscure little journals. About the best reception I ever got in presenting our thoughts was when one scientist in the audience said to me quietly after a talk, first exhaling a long sigh, "I guess I'm just not ready to understand this." The rest were not so kind.

Even so, our first summer's thoughts seemed to strike a chord with NASA, which began to support some research by its scientists at some of its eight research facilities on the characteristics of such a space station. But it was the engineers, not the biologists, who seemed to take the most interest. They chose to follow the other path that I had envisioned for this work from the beginning: They began to look at the problem as an engineering one—how many tons of this and that would have to be brought along so that this ecological life-support station would function.

The engineers issued a report. They compared an ecological life-support system with a conventional one. In a conventional space vehicle, everything was brought along. Water was obtained from fuel cells that "burned" hydrogen with oxygen and created electricity. The waste was pure water. Carbon dioxide was recycled directly through a chemical process. Food was dried and stored on board. It was like a long picnic rather than living at home on a farm. The weight of a conventional system depended on the length of the space voyage. Each day required the same weight of materials, so a week would require seven times the weight of supplies as one day, a year would require fifty-two times a week's supply, and so forth.

The engineers drew a graph showing the weight required versus the length of the voyage. The weight of the conventional system increased by the same amount for every day the voyage took. In contrast, the ecological life-support system weighed the same for any length of voyage. It was very heavy, and it was heavy from the beginning.

So there was a trade-off. As long as the weight of all the food, water, and gases necessary for life was less to carry than the weight of the ecological life-support system, it made more sense to use a conventional system. Picnicking in space made more sense for short trips than creating a homestead, just as it did down here on Earth. The crossover point—when it became more weight- and energy-efficient to use an ecological life-support system—was when the voyage was long enough so that all the picnic supplies required if one carried them, equaled the weight of the ecological self-contained system.

But the engineer also pointed out that you did not have to make everything. There could be a hybrid system. For example, it made sense to carry vitamins, since these weigh little but are hard to make and it would be difficult to maintain all the life forms necessary to make enough vitamins. The work continued for a number of years, well into the mid-1980s, and the space engineers explored this trade-off in considerable detail. How much stuff was worth bringing along versus making on the spacecraft? They considered many possibilities. They concluded that a compact source of food heavy in protein and fat would greatly reduce the weight of the ecological system and that the ideal food for this purpose was peanut butter. So the lesson of years of work that began when I came across Jerry O'Neill's article about flying doughnuts with ten thousand residents was: If you are going on a long space voyage and plan to have some picnics, don't forget the peanut butter.

This was presented in complete seriousness in the engineering re-

port. Reading it more than ten years after our first meetings in Snowmass, Colorado, I wondered once again whether my pursuit of knowledge about nature was really leading to anything productive and useful.
Once again, a large bureaucracy seemed to see things in a different light
than a small group who could let their imaginations soar—soar and possibly crash, that is.

Twenty-three

HOW THE COYOTES
OUTWITTED A BUREAUCRACY

ig predators seem to have always posed a dilemma for people. On the one hand, whether about lions, leopards, tigers, sharks, wolves, or sea lions, one prevailing attitude has been that the only good varmint is a dead varmint. On the other hand, people have had a certain respect, even admiration, for the skills of predators and for their cleverness.

The stealth and secretiveness of some of the big predators are legendary. The most impressive story about the ability of a big predator to keep out of sight is the story of a female mountain lion in Glacier National Park. The National Park Service staff began to notice mountain lion tracks right in the middle of main tourist areas of the park, amid the lodges and staff housing. The tracks appeared overnight and began to occur with such frequency that the rangers decided they had better track down the lion and move it far away. They started tracking and discovered that the lion was a female who had a den with a cub in it underneath the porch of one of the park employees.

Mountain lions are the only big cat in the world increasing in number today, and considering the ability of this female mountain lion to hide near people, no wonder. Their stealth and ability to fool the "expert" ranger illustrates one of the reasons they can survive and prevail. It also illustrates how easily wildlife experts can be fooled.

But of course the wiliest and smartest of all animals of North America seems to be the coyote, known to science as *Canis latrans*—that is, a member of the dog genus and, more to the point, capable of interbreeding with domestic dogs, the offspring of which are known as "coydogs." The coyote has been a major symbolic animal in the myths of North American Indians of the Southwest, the Great Plains, and California. He is known in these cultures as the magician and trickster, as well as a creator, sometimes as the agent who brought people fire, daylight, and arts. Since the European settlement of North America, he is known mostly as a varmint, a pest to be exterminated. The coyote is infamous for the difficulty of finding a way to kill it, and even when this has been successful here and there, the total number of coyotes only seems to increase and the range of the coyote to spread.

The coyote gets its name from the Aztec *coyotl*, indicative of its wide range—from Costa Rica north to Alaska. Once known as an animal of the American West, in recent decades it has moved eastward and is now found in many East Coast states. Originally associated with the great outdoors, it is found today within the suburbs of Los Angeles.

As suburbs spread around the United States, the interaction between people and coyotes most likely will increase. Not only is the coyote wily, it is fast, said to be able to make a dash toward a prey at forty miles an hour. One thing seems certain: These animals are experts at what they do, including their ability to outwit bureaucracies as well as predator-control people, naturalists, ecologists, ranchers, hunters, and just plain tourists.

For a very long time, predator control dominated government policy and pioneer practice in America. When the East Coast of America was still a colony of Great Britain, various communities enacted laws that provided bounties for the pelts of wolves. But as the number of wolves has declined, respect for them and their persistence has grown. Most big predators are either listed as endangered under the U.S. Endangered Species Act or are subject to special attempts to help them. As early as about 1912, the U.S. Bureau of Biological Survey was given the job of controlling predators out west. From that time to the mid-1970s, the activity continued. Aldo Leopold, later to become famous as one of the most important proponents of conservation of nature, began his career in predator control; his early writings were antipredator. The Bureau of Biological Survey was simply following the conventional wisdom of the time. It was soon to become the U.S. Fish and Wildlife Service and, over the years, developed the same kind of fixed thinking that happened at the Bonneville Power Administration.

During the time I was working on the salmon problems in the Pacific Northwest, I visited with Lee Talbot, one of the members of the panel of scientific experts that I had set up for that project. Lee was the one who had told Richard Needles about the social behavior of gazelles in Africa, which Needles then translated into the story about the social behavior of sperm whales.

We were standing at the Sommette Automobile Race Track in West Virginia, about an hour and a half's drive from Lee's home in McLean, Virginia, talking about the dilemma of big bureaucracies as Lee prepared his race car. Racing cars roared round the track, gears whining, exhausts thundering. Conversation stopped each time a car ran by the starting gate not far from where we stood next to Lee's Formula Ford. Lee had needed an extra hand and I had come along to help push and shove and hold things while he and his chief mechanic made subtle adjustments to the machine and checked its safety.

Lee was then and still is one of the world's experts on wildlife conservation and one of the most successful in getting things done. Lee had extensive experience in government. He worked in the Council of Environmental Quality under the Nixon and Carter administrations. He helped write the Endangered Species Act and the Marine Mammal Protection Act. He had worked in a hundred and twenty nations on the conservation of nature for the World Bank, and at one time was head of the International Union for the Conservation of Nature. He was well known as one of America's leading thinkers and actors for the environment.

In addition to our love of the outdoors and our work to help conserve nature, Lee and I shared an interest in modern technology and in fast-moving vehicles. Our friendship grew out of discussions about both. I flew small airplanes and Lee raced cars. Lee was a world champion race car driver, having won or come in second in famous international events like the cross-Africa rally. And for several decades, he was a consistent winner in the United States race car circuit. I had never even considered trying to fly in an airplane race, but Lee not only participated, he made a science out of understanding how to win—and he did win, most of the time. He knew how to get things done on the racetrack, too.

His mechanic was also an expert at keeping automobiles on the move. Nobody thought I was an expert about anything at the racetrack. I was just the equivalent of a hired hand, someone to go and fetch or to pass a tool or help drive the vehicle that pulled the race car.

As he prepared his Ford for the next race, Lee talked about his perceptions of big bureaucracies. He saw a historical pattern in the development of government agencies: "A new idea becomes popular. People become outraged over the way things have been going following old ideas," he said. "Elected politicians respond by establishing a new government agency. That agency creates a need for a new kind of expert, and the agency has money. Universities recognize the new public need and the new money. New university departments and schools are established. Universities train the professionals required by the new government agencies. The universities and the government agencies benefit. So do the new faculty. They get their research grants funded and research

papers published. Faculty, universities, and agencies come to have common vested interests in maintaining the status quo. The faculty establish
a set of ideas and methods. These become the standards in use by the
practitioners. The government agencies rely on the professionals from
the universities for advice, and the university and the faculty rely on the
government for funding. Practices, ideas, and approaches become fixed.
What had been a new idea becomes societal inertia." He stopped talking
while several cars roared by.

"Eventually, because people are imperfect and our ideas and forecasts
are never perfect," Lee continued, "the bad as well as the good of what
were social innovations become apparent. Citizens become upset by the
bad effects. A new outcry begins, politicians respond, and the cycle
starts again."

I said, "Perhaps experts in political science will find this explanation
incredibly naïve and simplistic, but for those of us working in the
trenches, trying to find real solutions to real problems, that analysis fits.
The way I put it is: You start with a good idea and you end up with a bureaucracy—just like Woody Guthrie and the Bonneville Power Administration."

"I'll tell you about my own experience in getting things done in spite
of the tendencies of bureaucracies to become fixed in their ideas," Lee
said. "Early in my career—in the summer of 1948—I worked as a state biologist in California at San Joaquin Experimental Range, run by the U.S.
Forest Service and the University of California at Davis. Although the
experimental range was on Forest Service land, the university did a lot
of the research there. This was my first acquaintance with a poison
known as 1080 that was used to control predators and rodents—control
in this case being a euphemism for kill. [It was the same poison used in
California that many had blamed for some of the deaths of condors
when those birds still flew in the wild.]

"The university was conducting experiments to see how far this poison, 1080, went through an ecosystem food chain and within the cycling of chemicals in an ecosystem," Lee continued. "In theory, such poisonous organic compounds were supposed to decompose in their first victim, but it seemed that 1080 could kill a target animal and then remain poisonous and kill any animal that fed upon the one it had just killed. If you put out seeds with 1080 in it, you killed birds—any seed-eating birds. If you put out grain with 1080 to kill rodents, you would also kill birds, including raptors like hawks and eagles that fed on the dead rodents, and scavengers like turkey vultures that ate the carcasses."[6]

From that summer experience, Lee became impressed with how dangerous this chemical was to an ecosystem, not just the target animals. In the years afterward, he saw the same food chain effects with other poisons put out to kill varmints, and he became concerned with the widespread use of any broad-spectrum poison, especially in the western states.

The predator-control experts, most of whom worked for the U.S. Fish and Wildlife Service or for state agencies and were known as "gopher-chokers," operated as "laws unto themselves, regardless of landowners," Lee said. "Over the years it became clear that predator control operations of the government were not under control, and the biologists and control officers involved seemed to have little idea what impact they were having," he continued.

At the time, poisoning predators and prairie dogs and other rodents was popular with some ranchers—including the more powerful rancher groups. "Like any bureaucracy, the Fish and Wildlife Service predator control people pretty soon lost view of the finer points of their mission," Lee said, "and simply turned it into a way of expanding their endeavors— the more they could kill, the bigger the body count, the better for them within their system—more jobs for them, more money to spend, more influence. So they went about their business without much of a scientific basis."

We strolled over to his car to watch what his mechanic was doing.

Lee stepped into his car transport trailer and brought out some tools that he handed to the mechanic. "All sorts of devices were tried on the coyote," Lee continued. "One was the coyote getter—a shotgun cartridge buried in the ground with just the top showing, a powder charge in the bottom of the cartridge and a charge of strychnine or 1080 at the top. Some kind of bait or a piece of cloth soaked in a coyote attractor sat on top of the cartridge, connected to it. When the cloth was pulled, the cartridge fired the poison into the puller's mouth." The intention was that a coyote would be the puller.

Another method was simply to throw poisoned bait, such as poisoned grain, out of airplanes—tons of poisons were released indiscriminately this way, especially for rodent control, he said.

Lee had waited for a number of years to do something about this. In the 1970s, when I was trying to learn how long a whale slept, how many leaves were on a tree, and was roaming forests of New Hampshire with my father-in-law and beginning to study wilderness and to try to help save endangered species, Lee took a more direct approach. He joined the staff of the newly created Council on Environmental Quality, a new part of the executive branch of the federal government formed during the Nixon administration. Several things came together about this time. First, the increase in public concern with the environment meant that it might be possible to educate the public about the role of predators in nature and the degree to which they were or were not varmints. Second, President Nixon recognized the public interest in the environment and believed that there could be political advantages for him to do good things about it. The question was: what things? His aides, Ehrlichman and Haldeman, were actively looking for politically visible good things to do for the environment. Third, the staff of CEQ was looking for effective, scientifically sound things to do for the environment.

At long last there seemed to be some possibility to do something to stop the use of 1080, but there was concern about the power of western stockmen and the employees of the Fish and Wildlife Service. "To get something done in our nation, you need a triggering event—something that grabs the public's attention and creates an opportunity for political action because the action will show that the politician is listening to and responding to the needs of the people," Lee told me. You had to wait and watch for the right event. It reminded me of the fox on Isle Royale watching the squirrels, waiting for just the right time for action.

"With 1080, the trigger was the day that a young boy shot himself in the face with a coyote getter," Lee continued. "He was on a camping trip out west with his family and came across a piece of cloth lying on the ground. A young boy's typical curiosity was aroused, and he pulled on the cloth. The shotgun cartridge exploded and fired 1080 into the boy's face. He was okay, but the event got a lot of media attention: a young boy shot with poison from a device set out by a federal government agency to kill coyotes."

Lee's mechanic motioned for us to help him push the Formula Ford to a better location for him to work underneath it. We helped him, continuing our conversation as we gently pushed the light vehicle.

"A second triggering event added to the situation: a series of eagle killings out west came to light," said Lee. "One of the federal predator-control people had been taking local ranchers up in an airplane so that they could shoot and kill eagles. This was in direct violation of two federal laws, one a special act to protect the eagle, America's symbolic bird, and the other the Endangered Species Act.

"They and the ranchers argued that eagles were a threat to livestock. Eagles were believed to be able to kill calves and newborn sheep, an unlikely event considering the relative size of an eagle and a calf, and a calf's mother."

This was a continuation of the persistent belief that the only good varmint was a dead varmint. "Even today in parts of the ranching west, this belief still holds," Lee said. "The intentional violation of two federal laws by zealous federal predator-control experts hit the national press, creating public concern. Major national environmental organizations jumped on this event."

Lee told me that he used the two triggering events and prepared a proposal for the White House that took into account public outrage over the injured young boy, the illegal killing of eagles, and long-term environmental concerns about the release of broad-spectrum poisons into the environment. Nixon saw this as an environmental issue that could create a positive image for him. Lee was able to promote the idea of eliminating 1080 and the uncontrollable weapons that involved this poison.

The question was how to put a plan into action, given strong political opposition in the West and from the ranchers' congressional representatives, as well as from the predator control people and their allies in the U.S. Department of Interior. Lee got together with the then assistant secretary of interior for fish, wildlife, and parks, who was also concerned about the unnecessary killing of predators and the unannounced spreading of poisons into the environment. It would be typical for a large federal bureaucracy to close ranks behind the predator-control people, Lee told me, but this assistant secretary was an exception.

The assistant secretary and Lee recognized that they needed an impeccable scientific rationale—the blessings of the right kinds of experts. They considered forming a National Academy of Sciences panel to examine predator control and the use of broad spectrum poisons. But they decided that this would take too long, losing the advantages of the triggering actions. Instead, they put together an expert committee of top authorities on predators and on the impact of predators on their ecosystems. Stanley Cain, a well-known scientist who had been Dean of the School of Natural Resources at the University of Michigan, headed the committee. "His credentials as an environmental scientist were impeccable," Lee explained.

Other members of the panel were Starker Leopold, Morris Horn-knocker (known as the world's top big-cat biologist), and Fred Wagner, a professor at Utah State University who had done much work with the livestock industry. They were asked to assess all available information about the impact of predator- and rodent-control operations in the West and report back with recommendations about what ought to be done. Lee made clear that the panel was completely independent and that Lee would not have any contact with the panel until the study was completed, so that it would be fair and impartial. The panel held hearings around the West.

The Cain committee, as it came to be known, reported its key findings: Vast amounts of poison were being applied indiscriminately in the West, including the tossing of them out of airplanes, with no control and no idea of the real impact. The committee had data on the killing of non-target species, including endangered species, as well as the role of the predator-control people in exterminating the wolf, grizzly bear, and mountain lion from most of their ranges.

The Cain committee provided an independent, scientifically impeccable report. The report stated that the "committee found that consistent poisons had been applied to range and forest lands without adequate knowledge of how they affect the ecology or if they actually prevent loss of livestock." It went on to note: "Large-scale use of poisons ... has unintentionally had effects on" nontarget animals. One finding was that it wasn't possible to identify the actual total impact of the indiscriminate release of poisons, but that since this release did not seem successful against coyotes, much was being done with no knowledge of effect even on the livestock that it was intended to save.

With this independent assessment, Lee wrote up an executive order for Nixon to issue. The arguments were presented to Nixon in a brief decision paper. (Nixon said he would not read anything longer than a page, and better only a paragraph.) In 1972, Nixon issued Executive Order Number 11643, titled "Environmental Safeguards on Activities for Animal Damage Control on Federal Lands." The important statement is the policy in Section 1:

It is the policy of the federal government to (1) restrict the use on federal lands of toxic substances for the purpose of killing preda- tory mammals or birds; (2) restrict the use on such lands of chem- ical toxicants which caused any secondary poisoning effects for the purpose of killing other mammals, birds or reptiles [this gets at the rodent control]; and (3) restrict use of both types of toxicants in any federal programs of mammal or bird damage control that may be authorized by law [this stops activities on federal land].

Lee began to get ready for the next race. He brought out his fire- retardant coveralls and started to get into them. He told me the report also stated that all mammal or bird programs should be conducted in a manner that contributes to the maintenance of environmental quality and to the conservation and protection, to the greatest degree possible, of the nation's wildlife resources, including predatory animals.

He reached for his helmet, telling me, "In the president's message on the environment in February 1972, Nixon said that Americans set high value on the preservation of wildlife."

He remembered a quote from that speech: "The old notion that 'the only good predator is a dead one' is no longer acceptable as we under- stand that even the animals and birds which sometimes prey on domes- ticated animals have their own value in maintaining the balance of nature."

The executive order barred the use of poisons for predator control on all federal lands with very few exceptions, and in all federal programs on public or private lands. EPA suspended and canceled poisons used in predator control.

Here was an example of a success story. How had Lee done it? The basic elements involved applying scientific knowledge. This required an excellent scientific assessment by a group of scientists who would be recognized as experts. But to make the whole thing work, one had to take advantage of crucial timing. This means that Lee had to understand the opposition and had to have worked out the plan of action in advance; then—when the opportunity arrived—he was able to put the plan into operation quickly, in this case getting White House approval and lining up as much scientific backing as possible.

It was a relief to hear of such a success story in conservation. Then again, no one has yet figured out how to control coyotes when they are pests—and sometimes they are, just as raccoons in my garbage cans were in Woodbridge, Connecticut. Success, like all things in the real world, has its limits. Because Nixon's action was an executive order rather than a congressional law, it could be reversed by subsequent presidents. And President Reagan did this—he brought back the use of 1080, and it became a problem with the condors in the 1980s.

As Lee climbed into his fireproof coveralls, sat down in his race car, and put his seat belt on, he reminded me that such successful action requires a triggering event and that you have to be ready to act—to have the scientific understanding available. His story also made clear that it wasn't easy to create policy that helped nature; it took a combination of applied science, scientists, and a lot of political savvy. It was the way Lee had succeeded in getting things done in government and it was the way he approached racing and winning races in his Formula Ford. His engine started. We put our fingers against our ears and watched him drive slowly out to the track.

Twenty-four

KILL THE SEA LIONS;
SAVE THE SALMON

All the mind's activity is easy if it is not subjected to reality.
—MARCEL PROUST,
REMEMBRANCE OF THINGS PAST: CITIES OF THE PLAIN

 y the late 1990s, I thought that surely things had improved. Ecology and environmentalism had become household words, perhaps even ho-hum words. Our 1970s misadventures with sperm whales and other creatures were long ago and far away. And there were lots more ecologists. The Ecological Society of America and the British Ecological Society had a combined membership of more than twelve thousand. Many subdisciplines had developed: paleoecology, restoration ecology, microbial ecology, evolutionary ecology. And then there had been the growth of environmental bureaucracies and government agencies, as well as the growth of environmental activist organizations. These bureaucracies and organizations employed experts in wildlife, land management, and many other fields. Environment had been a growth industry. Among the many issues, whales and other creatures that live in the ocean had gotten a lot of attention, as anyone who watched the Discovery Channel, the Planet Earth channel, and PBS's *Nature* knew.

At the beginning of the 1990s, I was involved in an Oregon project to examine the relative effects of forest practices on salmon—to do an objective scientific analysis, independent of ideologies and fairy tales. It was as part of this study that I met Lee Hill and learned from him about salmon and water flow.

Ever the optimist, I had readily accepted the opportunity to work on the salmon issue. Perhaps this time I can really make a difference, I had thought, even though the study of salmon was as far from my usual work as whales had been in the 1970s. Surely these fields—fisheries and whale biology—had advanced greatly in two decades. Why not? Most other sciences had made major advances in that time.

Soon after I began the study of salmon and forestry, I learned that there were four traditional culprits blamed for the demise of salmon: fishermen, for catching too many; foresters, for destroying their fresh-water habitats, especially where they spawn and where they spend their youth; engineers, for building dams; and sea lions and harbor seals, for eating too many. As part of our study, we searched for all the scientific evidence available about each culprit.

The hatred of sea lions by champions of Pacific salmon goes back a long way, I soon learned. In 1899, the president of the California State Fish Commissioners decided "merely to kill 10,000 sea lions," on the ground that they were eating so many salmon as to be "highly destruc-tive to the salmon fishery." He estimated that there were thirty thousand sea lions in California waters and offered a bounty for each one killed. At that time, local fishermen were quoted as saying "without qualifica-tion that sea lions feed extensively on salmon."[7]

This decision raised an interest among scientists all the way to the East Coast. Among them was C. Hart Merriam, director of the Biologi-cal Survey in Washington, D.C. (later to become the U.S. Fish and Wildlife Service). A few years before, when a similar claim had been made against fur seals, Merriam had traveled to California, found fur seals that had been shot for this crime, and cut open their stomachs. Examining "a large number of these animals," he wrote in *Science* magazine in 1901, he found to his "surprise" that "the great bulk of their food consisted of squid. In those seal stomachs were hundreds of squid beaks and pens," the hard and indigestible parts of those animals, "while in only a few instances were any traces of fish discovered."[8]

A professor at the University of Kansas, L. L. Dyche, also became curious about the sea lion threat to salmon. In the early fall of 1899, he traveled to the California coast and opened the stomachs of twenty-five sea lions killed by bounty hunters. The hunters told him that those twenty-five were especially voracious salmon eaters. Merriam wrote that Dyche opened the stomachs of eight of the sea lions in the presence of the fishermen who had shot them "because," the fishermen said, "they were feeding on salmon." The stomachs contained squid and octopus (including some pieces of the giant octopus), but no fish scales or bones. And these sea lions had been swimming in the midst of waters where fishermen were actively catching salmon.

"You can hardly imagine the surprised look on these fishermen's faces," Dyche wrote, when he cut open a dead sea lion and found "masses of squid meat." Dyche also went to islands off Carmel, California, where sea lions moved out of the water to sunbathe and mate. He dug through sea lion excrement—the animals were so abundant that their droppings made a foot-thick layer—looking for evidence of salmon, but found not a single fish bone or scale.

Here it was again: Ecologists trying to solve environmental problems often seemed to end up digging in shit. It was one thing I had learned over the years about the study of wildlife: It is hard to get away from digging your hands into scat, droppings, pies, manure, pellets, guano, and the many other names given to what comes out the back end of animals. Some of the best information has been obtained this way.

One of my best friends and closest colleagues in ecology over the years, Peter Jordan, an expert on moose, has spent forty years surveying the pellets and pies of moose in Isle Royale National Park as an index of the abundance of the animals. "Pellets" is the polite name for the hard little objects that moose drop in the winter when they are reduced to eating leaves of fir and twigs of other trees and shrubs. "Pies" is the term used for the mushy, watery piles the moose drop in the summer when they are feeding on water-filled green leaves of trees and shrubs in the forests, and water plants in ponds, streams, and harbors.

You may not believe this, but moose pellets can be ornamental. They are shiny, about two inches long, and have a regular shape, widest at the equator, curving smoothly toward the poles, like an elongated Earth. Indians string them on beads and sell the necklaces to white guys as presents for their girlfriends.

The pellets and pies stay around at least a year and don't move about much, so they are much easier to find and count than the moose themselves, who in spite of their size—up to a thousand pounds—are hard to see in the deep woods and dark ponds and bogs that they inhabit.

But back to the sea lions. Merriam, writing in 1901, acknowledged that sea lions in captivity would eat fish rather than starve, and that they sometimes bit and ate salmon in nets. But he concluded that adult salmon are just too big and fast for seals and sea lions to catch. Now and again they can take a bite of one and leave a scar.

Recently, we've seen striking exceptions to Merriam's observations of a century ago—man-made structures that make salmon easy prey for sea lions. The most notorious of these is Ballard Locks in Puget Sound in the state of Washington. There, a fish ladder allows salmon to climb up the mouth of a river where locks built to make ship passage possible would otherwise obstruct the return of the salmon to rivers to spawn. But the fish ladders create a cafeteria-style feeding arrangement for sea lions. Without even collecting a tray, sea lions and seals can swim over to the locks and pick off salmon as they struggle up the ladder. And there is no checkout counter.

Much effort has been spent trying to deter sea lions from this nasty habit. Sam Smyth, a Santa Barbara expert on marine mammals and a marine mammal tour guide who became a friend and colleague after I moved to that city, has had the contract to go up from Santa Barbara to Puget Sound, capture the criminally feeding sea lions, and transport them to the Channel Islands off the Santa Barbara coast. The bureaucratic plan for these misbehaving animals is that they will stay in their new home down south and feed from what nature intended them to eat, whatever that was. But the sea lions, not aware of federal regulations, simply swim back to Puget Sound and indulge themselves in the much easier task of taking salmon off the fish ladder. Some of these animals have made the round trip from Santa Barbara to Puget Sound many times. Although they are considered evil culprits, sea lions are not subjected to capital punishment for their crimes, and they live to hunt another day.

In spite of the examinations of sea lion and fur seal stomachs a century ago by Merriam and Dyche, people in the 1990s, including fishermen and some scientists, continued to blame these marine mammals as a major cause of the decline of salmon. But strangely, in spite of a century of controversy, we could not find a single well-executed scientific project that determined the exact take of salmon by sea lions and seals. In the 1970s, I had tried to find out how many hours a whale slept. Here I was with the much simpler and, you would think, more economically important question: How many salmon does a sea lion eat? And I could not find a good study that answered the question.

One graduate student who completed his thesis in 1988 came close. He did almost everything right, but he left one thing out: He measured the percentage of the stomach content of sea lions that was salmon (a very small percentage), but gave no weights for the stomach contents, so it was impossible to calculate the possible take in terms of weight or numbers of fish.

A more typical kind of study was done on the Rogue River—at whose mouth I had directed the meeting in Gold Beach where I first met

Lee Hill. The Rogue is one of the most famous salmon fishing rivers in Oregon, a river also well known for its white water rafting—part of America's Wild and Scenic Rivers system. It is a beautiful and much loved river. About the same time that I was making computer whales seek computer mates in a Woods Hole laboratory, local fishermen were complaining that sea lions were eating too many salmon on the Rogue. Two scientists did a study in response to the complaint. But they spent only fourteen and a quarter hours watching sea lions who came into the lower reaches of the river and only saw sea lions trying to feed on any-thing ninety-three times. Mostly the sea lions tried to catch sea lamprey. In only two of the ninety-three observations did they try to catch salmon, and these were young salmon swimming downstream. These were two *attempts* to catch salmon, not necessarily successful. The sci-entists reported no estimates of the size or weight of the salmon the sea lions sought. Nor was there any attempt to count the total number of salmon in that part of the Rogue River. So it was not possible to calcu-late, even if these attempts were successful, what effect this would have on the salmon population.

In spite of this lack of data, an early 1990s report sponsored by the Oregon timber industry asserted that marine mammals were a major cause of the decline of salmon and gave lots of numbers to "prove" this. They claimed that their numbers showed that sea lions were taking an amount of salmon equal to 85 percent of the commercial ocean harvest. But when we went through their calculations, we found many mistakes, some in the misuse of data, and others simply mistakes in arithmetic; the authors of the report simply added, subtracted, multiplied, or divided in-correctly. Curiously, every mistake erred on the side of making the take of salmon by sea lions appear larger. So we published all of this in our re-port, including a table showing their wrong arithmetic and our correc-tions of it. We held several press conferences and public meetings and presented the results to government agencies, to a committee of the Oregon legislature, to environmental groups, and to foresters and fishermen. I had expected that this would create considerable interest, perhaps even anger. It did create a lot of interest in the organization of salmon fishing guides, who tried their best to help us and to promote a scientif-ically sound study. But it did not interest the newspapers or most people, and the fishing guides' efforts to stimulate the research were unsuccess-ful. The guides tried to raise money from government agencies, but none were interested. I tell this story often, and recently one fisheries scien-tist told me that a graduate student had finally done it right, but I haven't had time to track that story down. I hope it is not another triple-canopy rain forest.

So in the end, in the 1990s, nothing happened in response to our analysis of sea lions and salmon. Nobody defended the wrong calcula-tions and the poor studies. There was no outrage among the environ-mental groups, the foresters, the fishermen, or the legislators. In fact, there was, in general, no comment at all. Here was a case where we ac-tually found out something—it is more accurate to say that we had found out that some important knowledge did not exist—and almost nobody cared. So much for success in solving environmental problems.

And in the year 2000, newspapers reported that fishermen in the Pa-cific Northwest were once again calling for the shooting of sea lions, still claiming that they were an important reason for the decline of salmon. Once again anger, plausibility, and the strength of the conviction in what must be true overwhelmed any attempt to find out what was actually go-ing on, in this case among sea lions, harbor seals, and salmon. It seemed to be an example of one of our human failings: Don't take the blame for something. Instead, find somebody else to blame, especially somebody who cannot talk back and who has a reputation, throughout a civiliza-tion's mythology and folklore, of being evil. Blaming nonhuman preda-tors for the demise or extinction of an animal or plant is common, not restricted to sea lions and salmon. As Lee Talbot knew so well when he

was trying to stop the use of the antipredator poison 1080, wolves, coyotes, and mountain lions are repeatedly blamed for serious damage to domestic livestock as well as endangered and threatened wildlife. It is
always easier to blame something that cannot argue back than to look at
the numbers.

So for a hundred years, fishermen, conservationists, and government
agencies have called for the killing of sea lions because they take too
many salmon—despite evidence to the contrary. And nobody seems to
care when this is pointed out. Not only are many experts still accepting
fairy tales and the equivalent of just-so stories as fact, but it doesn't seem
to matter to anybody else when they do. I was reminded of the case of
Brookhaven National Laboratory, which was conducting a study of one
hypothetical kind of pollution, radiation, while simultaneously chemically polluting the ground water—and not talking about it. I thought
again about Roger Atwood's mother-in-law in her wilderness. In Oregon and Washington, people's views of sea lions and salmon had much
in common with her thoughts. In both situations, imagination and wish
replaced observation and fact. We could not even say what Walter Burroughs, the gentlemanly New Hampshire drinker, said to me that winter
morning at the mailbox long ago: How were we doing? Not "All right for
our age and habits," if by those we might mean the improvement in the
study of life and its environment. How nice it would be to find Walter
Burroughs's kind of gentlemanly honesty and willingness to look at himself in the great salmon controversy of our time. Perhaps taking the situation with a little humor and less self-righteousness would help us face
ourselves and the reasons why we cannot open up our minds and hearts
to the knowledge available to us—and to use our hands and our heads to
gather the information that we need to know but would rather not.

Twenty-five

THE UN-IRRADIATED FOREST

round 1990, out of nostalgia, I took a trip to the East Coast and went back to the radioactive forest to see what had happened to it. I went out to the forest with Neal Tempel, the person who long ago had replaced Dave, the first technician I had worked with at Brookhaven National Laboratory. After Dave had left to go back to medical school, Neal and I had worked together for several years at the laboratory and had become good friends, keeping in touch over the years. He had great integrity and was a wonder with machinery, able to fix anything and able to quickly understand how instruments worked and what the purpose of the research was. Neal had worked at the laboratory continuously since I had left it and had risen to be in charge of facilities for the biology division of the laboratory. Neal was the opposite of Dave. He took the work seriously, he liked the goals of the project, and was enthusiastic about the work.

Neal took me around the abandoned radioactive forest. The instrument trailer was decaying and the entire forest appeared abandoned. The
effect of radiation on ecosystems clearly was not fashionable. The double fences were still intact, however, and the inner and outer gates still
locked. Neal opened them and we walked down the same path I had
treaded many times so many years before. Some of the signs marking the
number of meters from the source were still in place, if tilted slightly at
an angle.

Ironically, since the end of the radioactive forest experiment, one
kind of ecological research that has gained great popularity is called
"restoration ecology." This is the study of how ecosystems restore themselves; it is an attempt to understand enough of these processes so that
we can speed them up—in effect restore damaged lands and ecosystems.
There Neal and I stood, within one of the best examples of natural
restoration following a bizarrely novel pollutant. Nobody had done the
follow-up study, trying to answer the question: Will the radioactive forest recover from what it suffered and, if so, how? Neal knew of only one
ecologist other than myself who had had enough curiosity about this
recovery to visit the site. The other ecologist, teaching at a small local
university, had a student do a summer project—never published to my
knowledge—about the change in the forest after the cesium radioactive
source was removed.

Fifteen years after the end of the experiment, the forest at ground
zero was a surprise. Much of it remained treeless, but there were low
grasses growing near the source. The once well-machined container for
the radioactive cesium was now just an abandoned box, looking like a pioneer's vegetable cellar. We were surprised to discover that a few of the
oak trees that we had been certain were dead when the source was ra-
diating the site twenty hours a day had sprouted a few new leaves on
small twigs growing out from the bark. Clearly, the roots of the trees had
been protected from the radiation by the soil. They had survived years

of radiation and were able to regenerate a few stems, and on the stems a few leaves.

Off to one side was a dense stand of trembling aspen, a fairly unusual species on Long Island. It usually grows much farther north in much colder weather. But it seemed to be growing vigorously in the poor soil, almost bare sand without any organic matter. The radioactive forest, once seeming to be the height of modernity, of advanced technology applied to ecological science, looked liked a ghost town, of no interest to anyone.

About the same time that Neal and I wandered around the un-irradiated forest, word slowly leaked out that the administration of Brookhaven National Laboratory was buying up houses outside the laboratory site but near to it. An investigation revealed that the leaders of the laboratory had come to know that dangerous chemicals from the laboratory had gotten into the ground water and had spread beyond the laboratory site. They were trying to buy up all the affected properties before anyone learned about this and brought a lawsuit. The result was that the entire administration was fired and a new company brought in to run the laboratory. The graduate student's summer study of long before had been right, but no one had listened. The irony of the situation was a bitter one. The laboratory had planned for one kind of environmental disaster and supported research about it, meanwhile creating their own disaster, which they did not recognize—or chose not to recognize—at the time.

So, fifteen years later, standing at what used to be ground zero in what was once the radioactive forest, I had mixed feelings and thoughts. Had this experiment been a step forward for science? Why was the study of the forest's recovery not part of the original plan, and even if it was not, why, if restoration ecology had become the current fad, was there no interest in the most useful restoration experiment? Was ecology simply a series of fads and fashions, masquerading as "science"?

Since the time I had held a Geiger counter at ground zero in the radioactive forest, environmentalism had developed into big business, big

bureaucracies, and big nonprofit organizations. In the 1990s, environmental Beltway bandits around Washington, D.C., competed for a $120
million contract from the Agency for International Development to provide international assistance in solving environmental problems. Friends
with specialized expertise in various environmental fields served as professional legal experts and told me that price was never an object; the
lawyers would pay them whatever they asked, sometimes $400 or $500
an hour. Some of the major national environmental organizations, listed
as nonprofits, had CEOs who pulled down salaries of a half-million dollars a year.

After Heman Chase died in the mid-1980s, the old mill sank into a
slow decay. The flume needed replacement again, but there was nobody
to do it—that is, nobody with the love and the time to do it for no pay. I
visited the mill in 1995 and found a new kind of clutter overwhelming
the odd nineteenth-century tools that still hung in the dust in the back
room. Now, where once the planer, the drill press, and a power saw
had roared and the building rumbled with the lifelike power of the water-
driven turbine, a front seat from a 1960s van stood, its stuffing spilling
onto the floor and floating over the abandoned workbenches. A window that looked out onto the millpond, broken ten years before, had yet
to be repaired. My wiring job was still sound, but the power saw to
which I had so lovingly added a safety switch stood idle and dusty. The
villagers of East Alstead argued about what to do with the pond dam itself. Nobody seemed to want to take responsibility for it, but home-
owners along Warren's Pond did not want the lake level to fall. The
sense of local independence and local cooperation seemed to be vanishing.
Country gentleman Walter Burroughs no longer walked to the mailbox,
a little stiffly after a night at the bottle; he was peaceful and quiet in the
East Alstead cemetery. Clarence and Elsie Goodenow lay still too, no
longer visiting the cows in the midst of a goddamned fairyland, their
barn as idle as the mill.

Big science seemed to be paying off in many disciplines. The discovery of DNA had led to an industry of genetic engineering, capable of doing things to life that no one had dreamed possible. Solar energy photovoltaic cells had increased in efficiency by ten times, and wind and solar energy were becoming economically competitive with fossil fuels as a source of power. Waterpower, which had seemed so benign and environmentally sound, especially when Heman and I admired the workings of the mill, was now considered an environmental evil. Computer technology had created the richest man in the world, and I write this manuscript on a notebook device that stores everything I have ever written, along with the entire contents, pictures and text, of the *Encyclopedia Britannica*. When I give talks, this notebook projects the images that once existed on Kodachrome slides. The capabilites of the digital data collection devices in the Brookhaven Forest, the ones that filled a small trailer, are now captured within, in fact dwarfed by, devices that I can put in my pocket.

But big science seems yet to have succeeded in creating equivalent revolutions in ecology and in the study of nature. It seems to me that in some ways Heman was right, at least about trying to understand the connection between people and nature. Big was not better: Government agencies were not doing a better job with a dollar than Lee Hill in Oregon, the old-time fisherman who thought about salmon.

Instead, big engineering corporations, with few if any employees with a biological background but with knowledge of how to manage the logistics of large operations, had captured the bulk of federal money for international environmental projects, and a fair bit of the national projects as well. Nongovernmental environmental organizations that depend on membership funds needed to have continual crises and were generally driven by ideologies rather than facts. Government bureaucracies in charge of federal environmental policies had fallen victim to the traps that bedevil all big bureaucracies: driven by an overriding goal to stay in business and avoid making mistakes. The way one avoids making mistakes is to do nothing or to do nothing new.

Ideas about our relationship with nature have languished, confused by wish, want-to-be, and imaginary worlds that have never existed. My visit to the radioactive forest brought up all these difficulties and was disconcerting. Where were we going, if we were "going" anywhere?

Twenty-six

&

GETTING THERE IS HALF THE FUN;
OR,
LIFE ON MARS

Well, if things hadn't improved much down here on Earth in the way we were dealing with our forests, fish, and landscapes, surely, I thought, things must have improved with our space travel—way beyond the early ideas of Jerry O'Neill and his orbiting doughnut satellite holding ten thousand sailing souls. Space travel still fascinated me, even though I had stayed clear of NASA research for more than a decade, having decided that peanut butter in space stretched the limits of what I wanted to imagine.

But then in 1999, a Mars lander just disappeared after it reached the Martian atmosphere. The peculiar thing was not the loss of the Mars lander—such losses are to be expected as part of the development of a major new technology—but it turned out that the NASA team that designed and developed the lander decided not to include devices to measure the condition and behavior of the spacecraft as it passed into and through the Martian atmosphere. The NASA team had chosen not to

measure navigation, but instead to pack onto the spacecraft more equipment for measuring the Martian environment—its soils, rocks, water, and
atmosphere.

What we now consider ordinary travel by commercial airliner in-
volves continually monitoring each aircraft's location and performance.
This is part of the routine of flying down here on Earth. You would
think that kind of information would be all the more important with a
spacecraft going to another planet. Sending a machine to land on Mars is
still an experiment in itself. The process of travel is as important at this
stage in space exploration as the study of the Martian environment. In
this sense, getting there is half the subject—half the fun—and deserves
its full share of equipment, even if it does not seem so to the scientists
involved.

The failure to measure the right things is a common irony of our modern information age: although our ability to gather information increases
rapidly and scientists are awash in data, strangely, often the very things
that should be measured are not.

The failure to make the right kind of measurements was also an eerie
echo of events a century ago in the development of that other flying machine—the airplane. At the turn of the twentieth century, Wilbur and
Orville Wright were at first awed by the list of great minds who had at-
tempted to develop a flying machine. But they soon discovered that
those great thinkers had gathered almost no data during attempts to fly.
Their awe vanished, because the brothers understood the importance of
measurements. One of the triumphs of the Wright brothers was their
recognition of the importance of the right kind of data and their invention of ways to obtain those data and integrate them into a theory of
flight. As recounted by Martin Combs in *Kill Devil Hill: Discovering the
Secret of the Wright Brothers*, the Wright brothers used this approach to
invent aeronautical science and aeronautical engineering, the direct fore-
runners of space exploration.

Almost a century before the Wright brothers, President Thomas Jefferson sent Lewis and Clark to explore the American West—to find a way to the Pacific Ocean *and* to report about the condition of the climate, soils, geology, and life of the newly acquired lands—a mission similar to that of the Mars lander. Part of the success of that earlier expedition was that Lewis and Clark managed to measure the key things about their travel as well as things of interest about the environment. During their 1804–1806 expedition, they measured with incredible accuracy the length of the rivers and the distance they traveled; they counted the number of grizzly bears encountered; and reported in detail about the condition of the geology, climate, and soils.

Similarly, the Mars lander was intended to provide information about the environment of the red planet. It is the same kind of information that we, on our own planet, often lack—an irony of the information age. I have often described this as setting policy by plausibility—if it seems right, it must be true. The disappearance of the Mars lander seemed to me a spectacular failure that should bring home to us this strange irony of our times.

It seemed all the more ironic because about this time I was reading the classic book *The Art of War*, written about 400 B.C. by the Chinese Sun Tzu. I got a copy of this book on the recommendation of Jim Brown, the state forester of Oregon, who had become a friend and colleague as a result of the work I had started almost a decade before in Oregon on forests and salmon. Once in a while, I traveled to Portland on business, and sometimes Jim and I would get together for dinner and talk about what we were reading as well as what we were doing. At our most recent dinner, Jim had said that he had just read *The Art of War* at the suggestion of another friend. He said he found it insightful about life in general and about forestry in particular, considerably beyond the limits implied by the title. That intrigued me, so I bought a copy. Reading it one day, I came across a passage that took me by surprise.

16. Now the elements of the art of war are first, measurement of space; second, estimation of quantities; third, calculations; fourth, comparisons; and fifth, chances of victory.

17. Measurements of space are derived from the ground.

18. Quantities derive from measurement, figures from quantities, comparisons from figures, and victory from comparisons.[9]

We need only substitute "chances of success" in a project to study Martian or Earth's environment to apply his precepts to modern times.

Measurement of key factors is an old lesson, one that we ignore as much as forget. It is an expensive thing to forget, as the $165 million cost of the lander suggests. I thought it would be sad if the exploration of Mars, or human exploration of space in general, were shelved because of a failure to remember this old lesson. Perhaps there was more of a connection between space travel and ecology down on the Earth than I had assumed. It was becoming clear to me that the way we dealt with problems within our Earth nature was reflected in how we tried to explore other ecosystems—other possible homes—in our solar system.

Twenty-seven

THE GREAT BASEBALL BAT CRISIS

fter I moved to Santa Barbara, I began to make one or two trips each summer to Dodger Stadium in Los Angeles to watch big-league baseball. I had always been a Dodger fan, going back to their glory days in Brooklyn when I was a kid. Times had changed not only for the environment and people's concerns with nature, but also for the once Brooklyn Dodgers, who now played in a beautiful ballpark with palm trees greening the darkening sky as game time approached.

At one game there seemed to be an unusual number of broken bats, and I began to think about the supply of baseball bats. They were, and still are, made of white ash, the tree that was my favorite one to split because of its straight, tight grain—the same reason that its wood made such good bats. The best white ash for baseball bats comes from Pennsylvania, but white ash no doubt also grew south in the mountains near Louisville, Kentucky, where the famous Louisville Slugger baseball bats were made.

Who was taking care of the supply? White ash is a peculiar species, with an odd assortment of requirements in the eastern forests that are its home. I have found it mostly growing near or along streams, but it also seems to love fertile, well-watered and well-drained soils wherever they occur. Surveying with my father-in-law Heman Chase, I found white ash upland, away from streams, if the soil was especially rich. It was not a straightforward tree in its requirements. Its niche, as ecologists called it, was peculiar. Murray Buell and I, usually convinced we knew why a tree grew where it did, stopped a long time and looked in silence at a white ash that stood straight and tall in front of us, breaking the easy rules for what a tree was supposed to require. A tree like that might be difficult to grow.

I am one of those fans who loves to hear the solid *thwack* of ball against wooden bat, especially when it is a home run. I hate the loud *ping* of aluminum and, as an amateur player, hated the funny feel of an aluminum bat. It vibrated in my hands when I hit the ball as if it were one of those trick toy buzzers. Its shiny, light, and vibrating material seemed to be part of a child's toy game, not a thing of our national sport, something that one takes, should I say, seriously.

As bat after bat broke during the Dodger game, I began to think that a great baseball bat crisis might be looming just over the horizon, more serious than one strikeout in a game or an entire bad season by my favorite team. Sustainability of the supply of baseball bats was the issue. This was no hypothetical, academic issue to be discussed in the stacks of a library. It was the essence of the American way, by golly.

It was bad enough that aluminum had replaced wood in high school and college baseball. But unless a sustained source of the white ash was established, aluminum might be the material of major-league baseball's future. A depressing thought. If aluminum bats were going to be the future of baseball, our society, our civilization, was going downhill rapidly. Baseball, after all, is the game that symbolizes American democracy and teamwork. And it also symbolizes the connection between people and organic life. Baseball is a game of the products of living things: a leather glove, a leather-covered ball with twine in it, a white ash bat. Baseball could not exist in a world without oxygen produced by trees like white ash, algae, and photosynthetic bacteria. The ancient world before these kinds of life altered the Earth's atmosphere was not a place where one could play the game. Only in an oxygenated world could multicellular, three-dimensional, warm-blooded creatures run, hit, and field. Thinking about this connection between all of life and the game of baseball, I wondered what kind of sports those inhabitants of Jerry O'Neill's orbiting doughnut space station would play. It could not be real baseball, as a soaring home run four hundred feet to the center bleachers might go right through the skin of his space station. And in the thin air, a well-hit ball probably really could go a country mile, if that kind of space were available. Well, so much for the joys of that kind of life, I thought.

Not only is white ash a dream to split, but it also can be turned readily on a lathe to the smooth finish so familiar to anyone who has played baseball. And its straight grain and hard wood make it strong, resilient, and less likely to split than other woods. That is why white ash is the wood of choice.

When I got home from the game, I went to the library and to my local bookstore and looked into the facts about baseball bats. According to major league rules, a baseball bat can be no more than 42 inches long and $2\frac{3}{4}$ inches in diameter. As a rough estimate, a bat could be turned on a lathe from stock from a four-foot 4-by-4. That's four-tenths of a cubic foot of wood (the forester's common measure of wood quantity).

Searching a bookstore in Santa Barbara, I found George Will's wonderful book, *Men at Work: The Craft of Baseball*. In it he wrote that Hillerich & Bradsby—makers of the Louisville Slugger—sold 185,000 bats per year to the big leagues when there were twenty-six teams; at the same rate of use, the thirty teams that now made up the big leagues would need about 214,000 bats a year.

I started up my computer and ran the forest model I had developed so many years ago with colleagues in industry, playing the computer game of growing white ash in plantations and harvesting it at regular intervals. If trees are grown in plantations and harvested at regular intervals, called "rotation times," then the longer the time between cuts, the bigger the trees and the more bats can be made per acre. The computer forecasts suggested that an acre of good bottomland could produce about 100 cubic feet of baseball bat-worthy ash in forty years and 480 in sixty years. If the trees were cut every forty years, then 32,640 acres—51 square miles—would produce 214,000 bats a year. Each year, one-fortieth of the land would be cut. By the time the last of the acres in that 32,640 acres were cut, the first acre would be ready for another logging. If baseball can wait sixty years between cuts, only 10,240 acres will be needed per year.

To be on the safe side—in case of fire, hurricanes, insect outbreaks, disease, or just badly formed trees—it seemed a wise plan to leave a good margin for error and plant 20 percent more than the minimum. It would also be smart to ensure the game against the possibility of global warming, which might make Pennsylvania no longer the ideal habitat for white ash—that habitat might move hundreds of miles north. So one group of plantations would be where white ash grows well now (in Pennsylvania) and others would be where white ash might persist in the future (say, in New Hampshire). This would require twice the area. A total of 78,000 acres would be needed if the trees were cut when they were forty years old. Only 20,500 acres would be needed for a sixty-year rotation.

As I mentioned, white ash grows rapidly, but it likes good soils and is found naturally on fertile river floodplains where there is an ample supply of flowing water and well-drained soil with plenty of minerals. That land might not be easy to find. White ash also likes a lot of sun and grows best where there are not too many trees of other species nearby to compete with it for light.

Suppose you could buy good white-ash-growing land for a price of about $1,000 per acre. Then all the land you would need would cost $78 million for the forty-year rotation plan and $21 million for the sixty-year rotation plan. The second price is less than a recent multiyear contract signed by catcher Mike Piazza when he left the Dodgers for the Marlins, and then went on to the New York Mets, and less than that of Kevin Brown, a pitcher the Dodgers obtained from San Diego in 1999 and signed to a seven-year contract.

Of course, all these estimates are rough, back-of-the-scorecard calculations made when a game was especially slow. They could be too low or too high, but they give a general idea of costs; they're in the ballpark using available numbers. In addition to the cost of buying land, there are costs of planting and maintaining the trees—the work of professional foresters. But obtaining the land is the first step. Meanwhile, natural growth of white ash, as it occurs in the forests, would not seem to be keeping up with the demand.

Strangely, as best I have been able to find out, nobody is buying land and putting it into sustained plantations of white ash for baseball bats. Suppliers have expressed concern about the future supply of white ash for bats. For the price of the multiyear salary of one top ballplayer, big-league baseball could buy the land that would ensure us a future of wooden baseball bats. The choice is simple: one more big-salaried baseball star, or purchase the land to assure that wooden bats will be available indefinitely. For fans who love the hearty sound of wood against ball and detest the *ping* of an aluminum bat, the answer seems easy: Go with the wooden bats; persuade major-league baseball to take the first step to develop sustainable forestry for white ash. Even big-league baseball is affected by the need to conserve nature and to find a way to sustain the production of nature's resources. Well, if I couldn't solve all the environmental problems, as least the forest model that had seemed too strange to create years before might help our national sport become "sustainable."

Twenty-eight

THOREAU'S TRANSIT

[The new settlers of Pennsylvania take] little account of Natural History,
that science being here (as in other parts of the world) looked
upon as a mere trifle, and the pastime of fools.

—PETER KALM, 1750[10]

n the mid-1990s, I was canoeing in Maine with several companions, following the routes of Henry David Thoreau more than a century before us. Our plan was to make a film about Thoreau and nature. And I was writing a book to go along with the film.

It was early in the morning and we had just put our canoes into Oyster Pond on the west branch of the Penobscot River, several hours drive from Millinocket, Maine, beginning a two-day journey. We had canoed around the first bend, just out of sight of our put-in point, when we surprised a cow moose, or she surprised us. She moved quickly from the center of the stream up onto the bank, then turned and looked at us. Water dripped from her drooping, bearded jaw and she chewed on a stem of a water plant she had pulled from the bottom of the stream.

Although it was cloudy and beginning to drizzle, Ted Timreck, a professional filmmaker who had proposed the idea for a book and film about

Thoreau, pulled his movie camera from its watertight case and began filming, standing up in his canoe. Elaine Towers, a graduate student and experienced forester we had hired to help paddle, was in the back of Ted's canoe; she quickly maneuvered to shore. Ted jumped out and followed the cow moose as she alternately moved off slowly and stopped and looked at us—somewhat alarmed, possibly threateningly. Several times she pulled at a few leaves of a blueberry bush and seemed to calm slightly.

In another canoe, I sat and watched the goings-on with Parker Huber, our mentor and guide on the trip. Parker had walked and canoed every mile that Thoreau had traveled in the Maine woods. He had done this several times and written a wonderful book about it, a guide to Thoreau's travels with maps and photographs, titled *The Wildest Country*. He was a devotee of Thoreau and sought the same kinds of things from nature that Thoreau had sought more than a century before. Tall and lanky, Parker was comfortable with the outdoors, and was an excellent canoeist and companion. We became fast friends and talked about how Thoreau had approached the study of nature and how he had made contact with nature.

Journals of explorers and the first naturalists to travel in America have fascinated me since the 1970s. Part of the reason I had readily agreed to Ted's project was my fascination with the way North America looked before it had been changed by Western civilization and to rediscover that country now. But also, after several decades of trying to help solve environmental problems, I had to admit that I was curious how Thoreau and other early naturalists and explorers approached the problems they encountered in nature. Maybe there were some lessons from the past. Perhaps they even had some keys that might help me find better ways to solve environmental problems today.

Our destination that day was an inn, located on a promontory, called the Chesuncook Lake House, reachable only by boat. The promontory had held a place for people to stay since Thoreau canoed this route in 1849. During his time, the location of the inn was the home of an Ansel Smith, who cleared about a hundred acres, grew hay, and put up lumbermen and other travelers. It was, interestingly, the last building that Thoreau described on his trip in the Maine woods before he returned to civilization. Thoreau had canoed this route primarily to see and hunt moose with his Indian guide; he wanted to understand the entire Indian hunting process.[11] So our find of a moose was fortunate not only for Ted's filming, but also because we desired verisimilitude.

While we watched Ted following the moose, Parker and I talked about Thoreau. A light drizzle continued to fall, but it was a warm summer day, so we stretched out and chatted, ignoring the rain.

"Something that interested me about Thoreau's writing about the Maine woods, and I wanted to talk with you about," I said to Parker, "although he tells you a lot about the woods, he also spends a lot of time talking about what his Indian guide is doing and what his skills are. He admires the guide, who seems to have a sense of being in the place. This is different from many nature writers who just describe the countryside. Thoreau gives you his own feel for it."[12]

"Yes, yes, absolutely," Parker said. "He was very enamored of the Indians, wanted to learn their ways. And I think he gives us a genuine picture of the Indian, not romanticized—all the sides he saw of them. And there's another quality, too, in Thoreau's writings. It seems to me he's able to be present in the moment, I think that's what you were talking about. The experience itself is something he strives for—as much as he can—with *all* his senses."

All his senses. An interesting thought. I mused about it while Ted climbed back into his canoe and Elaine paddled them out from shore. The drizzle had let up, so Ted kept his camera out and ready. He was grinning broadly; we had thought we might have a chance to see a moose, but not within five minutes of putting in our canoes.

We continued on down the west branch of the Penobscot River, passing some mild white water that took our attention, but then the Penobscot smoothed out for a while, and Parker spoke up again.

"Being here. Canoeing. I think it gives us spiritual refreshment, the same kind of fulfillment it gave Henry," Parker said. "The same kind of discovery of yourself, as well as the natural world. There was an inner and outer world for Thoreau, and he struggled to understand both at the same time. Each helped give insight to the other. So I think just providing this kind of experience for ourselves—canoeing in the Maine woods as an example—it allows new things to enter. Things we were not necessarily aware of or expecting. Thoreau was open—he sought that kind of travel."

Spiritual refreshment. Discovery of yourself. An inner and outer world. Open to that kind of travel. Indeed Thoreau was. He had spent the last ten years of his life mainly in his hometown of Concord, Massachusetts, taking a four-hour walk every day to study the trees and flowers, birds and small mammals, to see how things changed with the seasons, and to be himself within the spiritual feelings of the outdoors.

Ted spoke up as he and Elaine passed us in their canoe.

"That's an interesting theme," he said. "What are the accidents that happen when you're traveling and how does that relate to your traveling and Thoreau's?"

"I think you're traveling for the unexpected and to just be here," Parker replied. "You don't really have to go anywhere. Matter of fact, we could just camp someplace and wait for it to happen, whatever that 'it' might be. You wouldn't really know what it would be, but something would happen, and we would certainly have a further understanding of ourselves in the natural world. It takes a great deal of patience to be a naturalist."

Another intriguing thought. Don't have to go anywhere. Can experience nature right where you are. As we passed spruce and fir on the slope above the Penobscot, I thought about that idea. "You don't really have to go anywhere." Thoreau had said it succinctly in one of his famous phrases: "I have traveled a good deal in Concord." He had traveled extensively through the Maine woods, a wild country in his day, as parts of it are in ours. But he discovered he did not need to travel far to discover himself or to discover nature. A patient naturalist. Not exactly a description of myself, but it did seem to fit Parker.

We passed a clearing in the woods, filled with blueberries and huckleberries.

"Do you feel you've been changed yourself by following his travels?" I asked Parker. "I mean, you've been everywhere he's ever been and you've been there several times."

"Well, that's a good question. I guess I ask myself each time I come back, what would I bring back from an experience like this?" Parker replied. "What would I try to incorporate into my life at home from this, is the question. And I don't always have an answer for that, but somehow there's a fulfillment that stays within me that helps to center me in the daily world, and so I'm able to retain that for a while. And by coming back, or the memory of it, to rejuvenate that feeling again, within."

Incorporate the experience into my life at home. A fulfillment that stays with you. Helps center you in the daily world. Rejuvenation. So it was the connection between the Maine woods and life at home that mattered to Parker, not the woods for themselves only. Searching for rejuvenation.

"There's something to do with the pace that we're moving," he said as we paddled. "We're moving ourselves naturally, with our own rhythm, so you come to see what that's like, and at times it's possible to repeat that rhythm in your daily life—it's coming back to solitude, to stillness, simplicity, silence, all these.

"So you can bring back a different sense of yourself because you've experienced that solitude and space and stillness. Silence was a great teacher for Thoreau, and in his walks he preferred to be alone because that was where a lot of insight came, in the silent spaces. But in Maine, here, where he was with the Indians, they're kind of guiding him into the natural world."

We grew silent ourselves and concentrated on paddling. It had been a magic moment, on the Penobscot River in the drizzling rain, immersed in nature's water and woods, listening to Parker talk about the reasons he kept coming back here. We passed a clearing in the woods, then passed by once again a dense forest of spruce, fir, and occasional birch. A bird flew above the trees, passing too quickly for us to identify it. Ducks swam on the river, keeping ahead of us, flying low over the water if we approached too close, then settling back down on the Penobscot to float patiently, waiting for us.

We came to a place on the shore where most of the trees were dead and standing. This caught our curiosity so we pulled our canoes out and looked over this patch of forest. We couldn't tell from our brief visit what had caused the dieback—maybe an insect outbreak or a flood-ing of the stream. But the forest was regenerating, with small spruce and fir growing under the dead trees. It was undergoing the process of succession that Thoreau himself had written about, a term that he was one of the first to use. It reminded me of Thoreau the scientist as well as, in our conversation of a while before, Thoreau the seeker, in search of nature and in search of himself: two curious sides to the same person.

We canoed the entire day, traveling eighteen miles and reaching the Chesuncook Lake House in late afternoon, with plenty of time to set up our tents and get ready for dinner, cooked by a French chef who, along with his American wife, owned the inn. There was more pleasant rain during the night, beating lightly on my tent, making the tent feel all the cozier. The storm's winds blew on the lake in front of the inn, and I could hear the waters lapping at the shore. I opened the tent flap, put my head out, and felt the rain and breezes on my face. All the senses. Center oneself in the daily world.

The next morning dawned bright and clear but with a stiff wind blowing straight onshore. Our plan was to canoe directly north across the lake and then up a channel to a take-out point. With the strong headwind, I was glad to have Parker in the back, the two of us canoeing together. We were buffeted by the bright waves, pushed back by the wind, warmed by the sun. All the senses.

Later in the fall, Parker, Ted, and I went to Cape Cod, where we continued to follow in Thoreau's footsteps. We visited a house in Wellfleet where Thoreau had spent a night in 1849. The house was still standing and in good shape, repaired and cared for by its present owner. In Thoreau's time, it was the home of John Young Newcomb, an old man in his late eighties whom Thoreau called the Wellfleet Oysterman.

"This is where Thoreau spent the night of October 11, 1849," Parker said. Thoreau had just come to the Cape, traveling by train, stagecoach, and then walking, accompanied by a friend.

"It was really his first night after the day of walking, where he had been walking along the beach. As it started to get later in the day and near dark, they decided to find a place to stay overnight, he and his walking companion. And they move inland into this area, which was then devoid of trees," Parker said.[13]

"There was no forest, so this was open land, and they found this house and knocked on the door and were able to spend the night here." I looked around. Although the house was in good shape, the forest was encroaching on it, pitch pines and small oaks growing where there had been a lawn, some close enough to the house to touch it when the wind blew their branches.

"They were invited in by John Newcomb," Parker continued. "He was a shipmaster who had done some oystering in his life and was eighty-eight years old when Thoreau arrived here. And his wife also was there; she was eighty-four. They were taken in by them, and Thoreau was very taken in by the Wellfleet Oysterman's stories. He had stories that went back to the American Revolution and, of course, Thoreau was interested about that, because growing up in Concord he had the Revolution as part of his background. If you look at the house, you will notice it has kind of a bowed roof here, and it was actually built by ship-builders."

Thoreau wrote about himself as a solitary person who didn't really like company or society very much, and greatly preferred to be by himself in nature. But that wasn't really always the case. Often he enjoyed people. This was especially true when he was on Cape Cod and visited the Oysterman. Many pages of his book on Cape Cod are devoted to the conversations that he had with the Oysterman. He liked down-to-earth people who were straightforward and who had contact with nature in one way or another—like Lee Hill, the old-time salmon fisherman, or Murray Buell or Heman Chase or Roger Atwood, the expert on African wildlife. He didn't like false society or artificiality, but when he met up with somebody who was a real character and had interesting things to say, he was very sociable and jovial.

I had brought along Thoreau's book *Cape Cod,* and opened it to the passage where he described his visit to the Oysterman:[14]

> The old oysterman had told us that many years ago he lost a "crit-tur" by her being mired in a swamp near the Atlantic side east of his house, and twenty years ago he lost the swamp itself entirely, but has since seen signs of it appearing on the beach. He also said that he had seen cedar stumps "as big as cart-wheels"(!) on the bottom of the bay, three miles off Billingsgate Point, when leaning over the side of his boat in pleasant weather, and that that was dry land not long ago.[15]

Trees once grew where now was ocean. According to the Oyster-man, the Cape was not stationary, the very land on the Cape moved. This idea intrigued Thoreau. He wrote that another person

> told us that a log canoe known to have been buried many years before on the Bay side at East harbor in Truro, where the Cape is extremely narrow, appeared at length on the Atlantic side, the Cape having rolled over it, and an old woman said—"Now, you see, it is true what I told you, that the Cape is moving."[16]

The idea that the land surface of the Earth was dynamic was novel at that time. More prevalent was the belief that life and its local environ-ment—oaks, bayberries, the soil, bedrock—were fixed, and created a per-manent, static setting. Thoreau did not simply accept the Oysterman's and the old woman's stories as true. As he continued to walk east and north on the Cape, he reached the Highland Light, where he talked with the lighthouse keeper.

A few days later, after Parker had returned home for other business, Ted and I arrived at the lighthouse and Ted set up his tripod and camera and began filming. The Highland Light was a landmark in Thoreau's time just as it is today, a classic white pillar rising above a white building on the edge of a picturesque dune, high above the beach and the water, fac-ing to the east, to the open Atlantic Ocean. It sits on the edge of an un-dulating landscape of dune grass, shrubs, small oaks, and pitch pines. It is a lonely but picturesque landscape. From the lighthouse, the dunes fall away steeply for a long distance. The edge of huge dunes provided us with a grand view of the shore. Far below, at the base of the dune on which the lighthouse sat, people strolled along the strand, tiny toy fig-ures to us.

The lighthouse was built in 1798 to provide one of the major lights to guide ships away from dangerous shoals along the coast of the Cape, and it performed that function during Thoreau's time. Today, the lighthouse is automated and no longer has a keeper.

To Thoreau, the lighthouse keeper was a different sort of expert from the Wellfleet Oysterman and the old woman. The lighthouse keeper had lived and worked at the lighthouse sixty years, a length of time that amazed Thoreau. Like the Wellfleet Oysterman and the old woman, the lighthouse keeper had observed the erosion and movement of the Cape. "According to the light-house keeper," Thoreau wrote in *Cape Cod*, "the Cape is wasting here on both sides, though most on the eastern"[17]—the eastern shore being the ocean side of the Cape at this location, confirming what he had heard from the Oysterman and the old woman.

But Thoreau did not leave things at that. He did not accept the opinion of the Oysterman, the old woman, or the lighthouse keeper without tests of his own. At the time, the Highland Light stood back from the edge of the dune a distance that Thoreau said was 330 feet. Having worked as a surveyor, Thoreau improvised a surveyor's transit so that he could make measurements. "I borrowed the plane and square, level and dividers, of a carpenter who was shingling a barn near by," Thoreau wrote, "and using one of those shingles made of a mast, contrived a rude sort of quadrant, with pins for sights and pivots, and got the angle of elevation of the Bank opposite the light-house, and with a couple of cod-lines the length of its slope, and so measured its height on the shingle."[18] He observed that the dune rose 110 feet "above its immediate base" and 123 feet above mean low tide. Next, he checked his measurements against those of other land surveyors. "Graham, who has carefully surveyed the extremity of the Cape, makes it one hundred and thirty feet," he wrote.[19]

Then he looked for signs of erosion. He found evidence of erosion about a half-mile south of the lighthouse, at the point of highest land in the vicinity. There along the dune he saw streams "trickling down it at intervals of two or three rods" that left erosional shapes like "steep

Gothic roofs fifty feet high or more," which were at one location "curiously eaten out in the form of a large semicircular crater."[20] Ted and I found similar features on the edge of the dunes near the lighthouse.

Still not content with the *opinion* of the lighthouse keeper nor the measurements he was able to take himself, he examined data kept by the lighthouse keeper. "We calculated, *from his data,* how soon the Cape would be quite worn away," Thoreau wrote.[21]

Thoreau made additional measurements when he returned to the Cape the following summer. "Between this October and June of the next year I found that the bank had lost about forty feet in one place, opposite the light-house," he wrote.[22] From these observations he concluded that the Cape was wearing away about six feet a year. But he was cautious about simple extrapolation and generalization from a few observations. "Any conclusion drawn from the observations of a few years or one generation only are likely to prove false," he wrote, "and the Cape may balk expectation by its durability."[23] This skepticism—even about one's own measurements and observations—is one of the important features of science and of scientists.

From the observations of local experts combined with his own investigations, Thoreau began to generalize about the dynamics of the geology of Cape Cod. "On the eastern side the sea appears to be everywhere encroaching on the land," he wrote. "Not only the land is undermined, and its ruins carried off by the currents, but the sand is blown from the beach directly up the steep bank."[24] From what the local expert, the Oysterman, told him, and from what he heard also from the keeper of the Highland Light, and also what he learned from his own measurements, Thoreau began to understand that nature was dynamic.[25] He actually acquired the data that he needed to move beyond what seemed plausible to reach a deeper understanding of nature. He was able to reach this scientific understanding and to apply science while at the same time he was the person who, like Parker Huber, visited and revisited the Maine

woods and many other New England landscapes, searching for and finding that inner and outer contact with nature, that sense of renewal, a
spiritual as well as a scientific understanding of nature, that Parker had
talked about while we were canoeing the Penobscot in the Maine woods
a few weeks before.

Why don't people find it simple to make such simple measurements
today, in our much more technological era, the information age? I wondered. Why were we always missing out on the obvious? Thoreau's
homemade transit on Cape Cod and his travels in the Maine woods
seemed, at least in Parker Huber's way of phrasing it, to have the seeds
of an answer. Maybe it was because he had that deeper feeling about nature and sought to experience it directly, inner and outer, rather than
treat it as a political issue viewed from afar, or an administrative task to
be completed as part of a bureaucrat's assignment. Thoreau's experiences, as he wrote about them in his journals and books, did seem to suggest a key to approaching the problems that surround us. His ideas and
feelings seem to fit in with what Sun Tzu had written more than two
thousand years ago about the importance of measurements, and what
Thomas Jefferson had written to Meriwether Lewis, and what Lewis
himself, along with Clark, had done in their travels across the American
West: measure, measure, measure, immerse oneself within nature.

I thought once again about what Parker Huber had said from the canoe on the Penobscot River: "I think you're traveling for the unexpected
and to just be here." A patient naturalist observing closely, seeking to
discover himself and nature, and making measurements.

THE ECOLOGY OF CANCER

y wife, Erene, sat in a hospital bed connected to a variety of strange machines by clear plastic tubes, each with an intravenous needle leading into her arm or leg. She was on chemotherapy treatment for non-Hodgkin's lymphoma, a cancer of the white blood cells. We were in Portland, Oregon, and I was working on salmon and forests; it was the winter of 1992.

"I keep wanting to find out what cancer is and what all these chemicals I'm getting do," she said, "but all I can find are these feel-good books—you know—how to be happy when you know you have cancer. I can't stand them."

Her doctor was talking with another patient a few beds away. In spite of cancer, Erene still looked as beautiful as ever, with her blond hair and blue eyes and her cheeks smooth and rosy.

The cancer had been a surprise, but she had not been feeling well for all the months we had been in Portland. The previous summer we had been on an ocean cruise. I was one of the lecturers for the passengers, and the cruise was about global environmental problems. We thought this would be a pleasant time for us, and it was, as the ship sailed along the European coast. Most of the other passengers were very old. One elderly man with a pleasant sense of humor said, "You two bring the average age on this cruise down by twenty years." But all during the two-week cruise, Erene said, "I don't feel right."

"What do you mean?" I asked.

"I don't know. It isn't as if I have some definite symptom. I just don't feel right."

Erene looked as gorgeous to me then as she always had, and she was rarely sick and even more rarely complained about it, so this was disturbing, but neither of us knew what to do. It wasn't like her, that was certain.

Then in the fall of 1992, we drove from Santa Barbara to Portland. When we stopped overnight in Ashland, Oregon, Erene said, "Feel my arm—it's lumpy. Under the skin." It was. Neither of us knew what that meant or what to do about it, so we continued on our way.

The day after we moved into a rented apartment in Portland, a pleasant place on the twenty-seventh floor with a view of Mount Hood, a strange thing happened. I flushed the toilet and the water that cascaded into the bowl was blood red. I flushed it again and again and the water kept coming that color. Part of me felt that this was an ominous sign, some warning from the gods. I couldn't help it, that was my human reaction. Another part of me thought there must be a rational, scientific explanation. And there was—we learned a few days later that a hot water heater some floors above had burst, and rusty water had gotten into the water supply. But I could not shake an eerie irrational feeling that something was wrong and that this was a sign that it was so. Here I was, the person always trying to find rational answers, responding in a completely different way. But then Erene had said for months that she hadn't felt right.

As the fall colors began to show and we explored the Columbia River Gorge, a national scenic area of rich Douglas fir forests alongside one of the world's greatest salmon rivers—one of the most grand and beautiful places we had been—Erene began to have stomach pains. At Thanksgiving, they became so intense that we went to a doctor who told her she had diverticulosis, an infection of the intestines. The doctor showed her a lump in her belly and said it was from this infection. Her pains grew worse and she was put on an antibiotic that was supposed to cure the infection and reduce the swelling. But the lump did not go away. She suffered a complete blockage of her intestines and was rushed to the emergency ward. The doctor said he had to operate and remove a short section of the intestines that probably had diseased pockets in it harboring the infection and causing the blockages. I dropped all my work and sat with Erene in the hospital. The operation, supposed to last forty-five minutes, went on for four hours, by which time I was a nervous wreck, pacing in the waiting room. Finally the surgeon, a wonderful, large teddy bear of a man, came into the waiting room and sat down with me. He was as nice a physician as one could ever hope to find.

"It's cancer," he said. "Strange—it was hard—and layered—like wood. Had to chop it out. That's what took so long."

"What kind?"

"Don't know. We'll know tomorrow or tonight."

He called me at home that evening. "It's a form of non-Hodgkin's lymphoma—in a lymph node that was pressing against her intestines. That's what caused the blockages. Once in a while if she moved wrong, the swollen lymph node would push against the intestine and close it. I'm sorry," he said. "Very sorry."

In the morning, before Erene had woken from the anesthetics, an oncologist came to see me at the hospital. "Everything's okay," he said. "The cancer was just in one isolated place—in one lymph node. We looked everywhere else. Took a biopsy of nearby lymph nodes. Nothing.

We washed out her intestinal cavity over and over. Nothing. She'll be fine. We'll just do a complete round of chemotherapy for insurance—just to be absolutely certain—but don't worry. That's why I'm here early—to tell her the good news so she doesn't get off to a bad start when the doctors tell her she has cancer."

This is so nice, I thought. The doctors could not be nicer.

Erene went through the round of chemotherapy—a week of daily plug-ins like the one we were sitting through that day—then three weeks recovery, then the same regime again. A week after each treatment she was sick to her stomach and exhausted. She lay in a lawn chair on our balcony and slept in the Oregon sun.

The doctors told her she was cured. We celebrated and returned to our regular life. Spring passed to summer, summer to fall. Each day I thanked God that Erene was alive and well. Then, just before Thanksgiving, Erene found lumps in her neck around her thyroid gland. Tests showed a rapid rise in a chemical associated with the breakdown of white blood cells. The cancer had come back. It had been lurking somewhere, waiting, like a spider in a trap.

So here we were, Erene and I, back in the Portland hospital while her body absorbed more of the chemotherapy chemicals. As best as I could understand it, these chemicals were used in a race between normal and cancerous cells—kill the cancer cells faster than the chemicals killed the normal ones. A little poison to cure a big one. Her doctor came over to her bedside. "How're you doing?" he asked. He was a young, trim, and very pleasant man with an open, helpful manner.

"I want to ask you a question," Erene said.

"Shoot."

"What are all these chemicals doing, anyway?"

"What are they doing chemically?"

"Yes."

"Well, that's easy. You're getting seven different chemicals, but they all do the same thing. They put a methyl group on guanine in the DNA and that shuts down cell division and kills the cancer cells."

"All do the same thing?" Erene asked.

"Yep. That's right."

"What does Taxol do?" Taxol was paclitaxel—a recently discovered chemical that had been found in western yew trees and that seemed to be effective in fighting certain kinds of cancer.

"Taxol? Completely different. It affects the structure of the cell. There are these little straight rods—microtubules—that are always being made and taken apart. Taxol prevents a cell from taking them apart. The cell fills up with the little rods and self-destructs."

"Completely different way of acting."

"Right."

"And Taxol wasn't invented by scientists; it was discovered in a tree."

"Right."

"So the people designing chemotherapy drugs were thinking about what to do in one way; nature had some other ways."

"Right."

The chemotherapy did not help. Erene's cancer had become resistant to it, the doctors said; this often happened if a cancer came back.

Erene grew worse. Her physicians suggested that she go to the Stanford Medical Center for treatment because it was one of the top centers for research on lymphoma. We drove there from Portland. Although Erene was weak and continuously nauseated, she remained cheerful and kept her sense of humor. Along the way, we stopped at a McDonald's because the frozen yogurt they sold settled her stomach. As we sat inside, we looked through the window at other people coming and going.

"Look," Erene said and began to laugh. One after another, each of the people entering the building seemed to be limping. We watched for a while and it did seem that most of America's leg injuries were turning up that day for fast food.

"I'll go to use the rest room. Meet you at the car," she said.

I stood outside in the soft sunlight. Erene emerged from McDonald's, looked around to make sure nobody was watching her but me, grinned broadly, and walked to the car with an exaggerated limp, like Igor in the movie *Young Frankenstein*. Although ill to death herself, she could laugh with other people about human shortcomings.

Not long after, Erene lost the battle with cancer. She was calm and brave to the end. As she lay dying, one of the medical staff came to me and said, "Even now, she looks beautiful. Look at the color in her cheeks. She's at peace. She has accepted this." She did indeed look like a beautiful princess simply resting.

She was buried in Santa Barbara's cemetery, on a bluff over the coast with a view of the ocean and the Channel Islands to the south and the Santa Barbara mountains behind, views that she had come to love.

After the funeral, I spent a while with my son and his family in Berkeley, California, not doing much. I would drive to Golden Gate Park, put out a blanket, lie down, and look at the trees. I wasn't wondering how many leaves were on them. I drove by San Francisco Bay, but I did not wonder how long the whales out beyond in the Pacific slept.

One day my longtime friend and colleague, Harold Morowitz, a biophysicist, came to visit me. We went out to a coffeehouse and sat in the Berkeley sun talking. For a while we talked about Erene. Then Harold told me what he was doing—working on the origin of life. "The standard theory—the organic soup theory of the origin of life—it violates the second law of thermodynamics," he said, "but that doesn't stop most of the scientists working on the origin of life from continuing to pursue that theory. They approve each other's papers, approve each other's grant proposals. They live happily with their old paradigm, even though it's impossible." So, ecology wasn't the only science that liked its old ways, its own beliefs.

I thought for a while. Thinking quickly wasn't my mood. Harold asked me about my work on salmon. I told him about the history of killing sea lions in the belief that these animals were eating too many salmon. I told him the story about how the evidence did not support that idea, but people kept suggesting it anyway. The old beliefs kept hanging on. "Something like what you just told me, Harold. The fisheries scientists review each other's papers and grant proposals. A former student of mine, who works for the California state department on fisheries, told me the other day that he had just been at a research meeting of top fisheries scientists. He said they're still using the old ideas, not just in Oregon but everywhere," I told Harold. "The student told me, 'They don't care whether their ideas have any relation to the real world or not.'

"They just are happy to continue their careers," I added.

"Tell me about the cancer," Harold said.

I told him the story about the seven chemicals all doing the same thing. Harold sat upright. "You know, Dan, you've just told me that you and I have to run a workshop on cancer. My wife's always telling me, 'Harold, why are you studying something that happened three and a half billion years ago—the origin of life? Why don't you work on something that matters, like finding a cure for cancer?'

"Look," he said, "we just talked about three scientific fields in which the old paradigms hang on and nobody cares. Meanwhile, the sciences and their applications suffer. It's one thing if it's about the origin of life or fisheries, but it's another thing if it's cancer.

"What we need to do," Harold continued, "is to get the best cancer researchers in the country in the same room with some of the best biologists who are not cancer researchers—the widest range we can think of—and just let them brainstorm. Get the cancer researchers to open up their way of thinking."

It seemed a great idea, especially as a memorial to Erene and as a way I could deal with what had happened to her, turn negative to positive. Use Erene's question to help the people doing research on cancer. I spoke with Erene's physicians to see if they thought the idea had any merit. They put me in touch with Tom Frei from Boston's Children's Hospital and Jay Freireich from the Anderson Center in Houston. These two had invented chemotherapy—cured the first person ever cured of cancer with chemicals, I was told. My doctors got in touch with them and tried out the idea of the workshop. They were enthusiastic. They said they were stuck for new ideas and welcomed new ways of thinking.

Harold found funding for the workshop, and he and I brainstormed about whom to invite. We were surprised and pleased to find the best minds in the country in biology and cancer research were delighted to come. Harold and I worked out a plan for how the workshop should run: Have a few cancer researchers talk about what they did; let the biologists listen; then just open things up for discussion—make it clear that there were no stupid questions, that no one was required to write a paper from this, that the whole idea was to mix minds and see what came of it.

The next spring, Harold and I sat in his university office the night before the workshop was to begin. We wondered if this could work at all. Putting total strangers not lacking in ego in the same room with nothing asked of them but to talk.

"You know, Harold," I said, "this is a complete experiment. It could be a complete flop. And my field, ecology, doesn't have much to offer. I have one thing to contribute. The way I understand it, the underlying idea about chemotherapy is that it is like hunting all the individuals of a species to try to kill them and cause extinction of the species that way. The cancer in your body, it's like a species; the cancer doctors are like the hunters. But, popular notions aside, that turns out to be a difficult way to cause the extinction of any species. It rarely works, and it takes lots of time and energy. A much easier way is to damage or destroy the habitat." I thought about salmon on the Columbia River and Oregon's other rivers. "For instance, in Oregon, if you own the water rights to a stream, you have the right to build a temporary dam across it during a drought. Do that for three years and bang, there goes a run of salmon. Extinct. And all you had to do was bulldoze a little soil across the stream. Easy. Meanwhile, trying to hunt down and kill every salmon is hard. It takes a lot of time and money, and it's unlikely to work.

"Another example—around the turn of the twentieth century, the elephant seal was hunted for its fur, to the point that there were few left. Some said about a dozen. The story is told that the British Museum sent out an expedition to try to find those dozen and shoot them. The idea was that it was better to be shot and stuffed and exhibited in the museum than to die an ignominious death out on the lonely California coast. But they couldn't find the last dozen—couldn't hunt the species to extinction. Today, there are more than a hundred thousand. See, if you think of cancer cells as something like the elephant seals, and the chemotherapy as the hunters' weapons, then you always miss a few and they come back. It's like what happened to Erene.

"But that's about it for the ecology of cancer. That's all I've got to say. Your field, biophysics and biochemistry—that's much more to the point—so I hope you'll take the lead in the discussion. I'll do my best to help the discussion along, but don't expect much from ecology."

We agreed that this workshop was a complete experiment and went into the meeting expecting the unexpected. We were amazed to find that the discussion worked. The cancer researchers were great—they were open to new ideas and, unlike so many scientists I had worked with, did not mind when the other scientists around the table basically told them that what they were doing was totally wrong. New ideas popped up. Contrary to my expectations, the cancer researchers even found the ecology metaphor eye-opening: Ruin the habitat of the cancer. I discovered that there actually was research going on with that goal. Before our workshop, that work had seemed strange to most of the cancer researchers. The scientists talked about whether cancer was a whole-body disease or a disease, according to the traditional model, that began with one bad cell.

At the end of the meeting, as people were leaving, Tom Frei stopped to chat. "I want to tell you that this is the best meeting I've ever been to in my forty years of research," he said. Quite to our surprise, the work-shop idea, with its open approach to discussion, had worked.

The workshops became annual events. I discovered that Harold's and my larger views helped—mine coming about because ecology forced you to think about the big picture and maybe also because of the strange combination of things I had done, from trying to learn how many hours a whale slept to studying elephants and what they ate and drank. Harold's seemed to come about from his sheer mental brilliance.

At the subsequent workshops, people who otherwise never would have met began to plan research projects. We invited a geologist to tell us about the history of life, and then discussed what kinds of life suf-fered from cancer and what did not. We talked about the evolution of cancer. One speaker told us that only two animals are known to get prostate cancer—men and dogs. Cat's don't, lions don't, wolves don't, elephants don't. We discussed why this might be.

We spent one workshop on the ecology of cancer, looking at it from the point of view of how the human body functioned as a whole system and looking at cancer as an event in evolution. Susan Mazer, an ecolo-gist who studied population genetics, listened to the cancer researchers describe their work. They said that they bred strains of mice so that the mice were very susceptible to cancer. That way, they could get results quickly—within three to five years, the average time of research grants.

"You're throwing out the answer," Susan said.

"What do you mean?" the cancer scientists asked.

"You're looking for how a body can become resistant to cancer and you're throwing out the mice that have that kind of resistance. You need to do the opposite experiment. Breed a strain of mice resistant to cancer. Throw out the mice that get cancer easily. Then study that resistant strain to see why the mice do not get cancer easily."

Susan's idea came from thinking like an ecologist and taking a big-picture point of view.

Well, I thought after the fourth annual workshop, maybe something had come out of my search to answer all those childlike questions, and all the travel, and all the people close to the Earth I had met. Maybe, just maybe, this lifetime of work had given me and my colleagues an ability to think broadly, and maybe, just maybe, it was having, in some small way, an effect on thinking about cancer. Erene's death had created a lot of life in the lively discussions that took place at the workshops in her honor, coming out of a simple question she had asked and the simple answer she had gotten. It was one of the most positive experiences of the years I had spent in science.

1. The facts about Maggie's Bend come from my friend and colleague Dick Pfilf, U.S. Forest Service (retired) and former head of many national forests, who worked for a while at the Clearwater National Forest. (It is possible that I have misheard or misremembered some of what Dick told me, and I take full responsibility for the presentation of his facts.)

2. Some of the scientific results of the research in the irradiated forest can be found in the following:

    D. B. Botkin, G. M. Woodwell, and N. Tempel, "Forest Productivity Estimated from Carbon Dioxide Uptake," *Ecology* 51 (1970): 1057–60; and G. M. Woodwell and D. B. Botkin, "Metabolism of Terrestrial Ecosystem by Gas Exchange Techniques: The Brookhaven Approach," in D. E. Reichle, ed., *Analysis of Temperate Forest Ecosystems* (New York: Springer-Verlag, 1970), pp. 73–85.

3. Some of the scientific results of this work can be found in:

    D. B. Botkin, D. S. Schimel, L. S. Wu, and W. S. Little, "Some Comments on the Density Dependent Factors in Sperm Whale Populations," in *Annual Proceedings of the International Whaling Commission, Rep. Int. Whale Commission*, Special Issue 2 (1980), pp. 83–88.

4. David White's formal statement about this research project was sent to me as an email:

Dear Dan,

Good to hear from you. As usual I am struggling to maintain a research program.

Should your dog drink from the toilet bowl? This is a more serious problem than it may appear, as in many areas of the world, fresh potable water is in very short supply. Proposals for dual water systems in which gray water is separated from potable water and used in toilets, etc. A hue and cry erupted as dog and cat lovers worried about the health of their pets who regularly drink from the ever-convenient toilet bowl.

It turns out there are two major types of Gram-negative water bacteria. The *Pseudomonas*-type bacteria found in potable water systems are opportunistic pathogens and the really bad guys, like most human serious pathogens (plague, enteric pathogens like salmonella and *Escherichia coli*). Gram organisms contain lipopolysaccharide (LPS) on their outer coat. Carbohydrates cover the surface and they are held in by lipid A, which glues the LPS to the bacterial greasy membrane. The lipid A contains fatty acids, which become extractable in organic solvents after mild acid hydrolysis.[1] Water biofilm organisms like *Pseudomonas* have 3-OH 10:0 and 3-OH 12:0 as the fatty acid components of LPS-lipid A. The enteric (fecal derived) bacteria and most Gram-negative human pathogens contain 3-OH 14:0.[2] Even the most fastidious toilet bowls cleaned daily contain a biofilm at the water-air interface. You can't feel it or see it, but it is there. Which type of bacteria does it contain—fecal- or water-based bacteria? In work supported by the National Water Research Foundation contract WQ1 669 524 94, human feces contained in mol% ratios of 7

[1] J. H. Parker, G. A. Smith, H. L. Fredrickson, J. R. Vestal, and D. C. White, "Sensitive Assay, Based on Hydroxy-fatty Acids from Lipopolysaccharide lipid A for Gram-Negative Bacteria in Sediments," Applied Environmental Microbiology 44 (1982): 1170–77.

[2] D. C. White, J. O. Stair, and D. B. Ringelberg, "Quantitative Comparisons of In Situ Microbial Biodiversity by Signature Biomarker Analysis," Journal of Industrial Microbiology 17 (1996): 185–196.

(0.6)/19 (4) 3 OH 10 + 12/3 OH 14:0. The toilet biofilm contained 72 (30)*/19 (4) of 3 OH 10 + 12/3 OH 14:0 LPS fatty
acids.

So the toilet contains pathogens to water organisms to enter-
ics of 0.37 compared to 3.8 in the feces. It is safe. High flush and
low flush did not make much difference except that the higher
the flush rate the healthier the bacteria in the biofilm.

Thanks very much.

Sincerely,

David C. White, M.D., Ph.D.

5. William Derham, *Physico-Theology; or, A Demonstration of the Being and
   Attributes of God, from His Work of Creation* (London: A. Strahan, et al.,
   1798).

6. The story about Lee Talbot's experiences in stopping the use of the poison 1080 against coyotes is based on interviews with him over several
   years, but primarily on July 27, 2001.

7. The discussion of sea lions as predators of salmon is taken from:
   D. B. Botkin, K. Cummins, T. Dunne, H. Regier, M. J. Sobel, and L.
   M. Talbot, *Status and Future of Salmon of Western Oregon and Northern
   California: Findings and Options* (Santa Barbara, Calif.: Center for the
   Study of the Environment, 1995).

8. C. Hart Merriam, *Science Magazine* in an article published in 1901.

9. Sun Tzu, *The Art of War* (New York: Oxford University Press, 1971),
   p. 20.

10. Peter Kalm, *Travels in North America: The America of 1750*, trans. A. B.
    Benson (New York: Dover, 1963).

---

*Mean SD.

11. J. Parker Huber, author of *The Wildest Country: A Guide to Thoreau's
    Maine* and expert on Thoreau's travels, in personal communication dur-
    ing this canoe trip. Undated publication.

12. The conversations in "Thoreau's Transit" are direct transcriptions from
    the soundtrack of film taken by Ted Timreck. Film in author's private
    collection.

13. Transcript of film made by Ted Timreck, #02:00:39:02–02:03:11:28. Film
    in author's private collection.

14. This next section is directly reprinted from D. B. Botkin, *No Man's
    Garden: Thoreau and a New Vision for Civilization and Nature* (Washing-
    ton, D.C.: Island Press, 2001), pp. 103–107.

15. Quoted passages are from the Writings of Henry David Thoreau, *Cape
    Cod,* J. J. Moldenhauer, (ed.), pp. 118–121.

16. Botkin, *No Man's Garden.*

17. Ibid.

18. Ibid.

19. Ibid.

20. Ibid.

21. Ibid.

22. Ibid.

23. Ibid.

24. Ibid.

25. Ibid.

Daniel B. Botkin is a scientist who studies life from a planetary perspective, a biologist who has helped solve major environmental issues, and a writer about nature. A frequent public speaker, Botkin brings an unusual perspective to his subject. Well known for his scientific contributions in ecology and environment, he has also worked as a professional journalist and has degrees in physics, biology, and literature. His books and lectures show how our cultural legacy often dominates what we believe to be scientific solutions. He discusses the roles of scientists, businessmen, stakeholders, and government agencies in new approaches to environmental issues. He uses historical accounts by Lewis and Clark and Henry David Thoreau to discuss the character of nature and the relationship between people and nature.

Botkin is known widely for his books about the idea of nature and the implications of modern science for our ideas about environment. His most recent book, published in January 2001, is *No Man's Garden: Thoreau and a New Vision for Civilization and Nature* (Island Press). His most influential book, *Discordant Harmonies: A New Ecology for the Twenty-first Century* (Oxford University Press), is helping change the way the citizens, agencies, and corporations view environmental issues. *Our Natural History: The Lessons of Lewis and Clark* (Putnam) uses the adventures of those explorers to explain what the American West was really like before it was changed by European settlement: a land of natural changes and challenges. He is currently editor of *The Atlas of Global Change* to be published by Oxford University Press in 2004.

Within environmental sciences, Botkin is best known for the development of the first successful computer simulation in ecology, a computer model of forest growth that has developed into a subdiscipline in this field, with more than fifty versions in use worldwide. Botkin has also been a pioneer in the study of ecosystems and wilderness. He has directed research on wilderness and natural parks in many parts of the world, from the Serengeti Plains of Africa to the forests of Siberia and the wildernesses of the Boundary Waters Canoe Area of Minnesota and Isle Royale National Park. His wide-ranging research includes studies of sandhill and whooping cranes, salmon and bowhead whales, and moose and African elephants.

A leader in the application of advanced technologies to ecological science, he was one of the first to apply satellite remote sensing to the study of forests, and the use of computer-based geographic information systems to environmental issues. He has helped develop major national programs in ecology, including the National Science Foundation's Long-term Ecological Research Program and NASA's Mission to Earth.

Botkin is also a leader in the application of environmental sciences and in attempts to solve complex environmental problems. Under contract with the Department of Defense, he recently completed a project to develop ecosystem management on the nation's military bases. Under a bill passed by the Oregon Legislature, he directed a three-year study concerning effects of forest practices on salmon and their habitats in twenty-six rivers of western Oregon and northern California. The project showed new ways to promote a recreational economy while conserving salmon, and to forecast salmon returns three years in advance. He directed a study of effects of water diversion on Mono Lake, California, under a special bill passed by the California legislature, a study that led to major changes in government policy for that lake.

He is the 1995 recipient of the Fernow Award for Outstanding Contributions in International Forestry, given by American Forests and the German Forestry Association. In 1995, he was elected to the Environmental Hall of Fame, housed at California Polytechnic Institute, Pomona, California.

He is the 1991 winner of the $50,000 first prize of the Mitchell International Prize for Sustainable Development. He has been a fellow at the Rockefeller Bellagio Institute in Italy and the Woodrow Wilson International Center for Scholars, Washington, D.C. He is a fellow of the American Association for the Advancement of Sciences, and a member of the Cosmos Club of Washington, D.C.

Botkin is currently chairman of the North American division of the Sustainable Use Program of the International Center for the Conservation of Nature (Geneva and Washington, D.C.). He has advised the World Bank about tropical forests, biological diversity, and sustainability; the Rockefeller Foundation about global environmental issues; and the government of Taiwan about approaches to solving environmental problems, development of nature preserves, and devising data systems for environmental monitoring. He served as the primary adviser to the National Geographic Society for their centennial edition map on "The Endangered Earth." He has served on a state of California scientific advisory panel concerning the recovery of the California condor, and the scientific advisory panel for the U.S. Marine Mammal Commission.

He has been an ecological expert lecturer and study leader on three Smithsonian Institution–sponsored cruises, including one on the Columbia River concerning the Lewis and Clark expedition; taught field courses in ecosystem ecology at the University of Michigan Biological Station; and taken college students on wilderness and natural history trips in New England, the mid-Atlantic states, Michigan, and California. For four years he lectured as part of training courses for National Park Rangers on ecology and natural history.

His other books include *The Blue Planet* (John Wiley), a textbook on global change; *Environmental Science: Earth as a Living Planet* (John Wiley; 4th edition, August 2002); *Forest Dynamics: An Ecological Model* (Oxford University Press); *Changing the Global Environment: Perspectives on Human Involvement* (Academic Press); and *Forest Succession: Concepts and Applications* (Springer-Verlag). He is the author of numerous scientific papers in ecology, and newspaper and magazine articles about environmental issues.

Botkin is Research Professor, Department of Ecology, Evolution and

Marine Biology, University of California, Santa Barbara, and President of the Center for the Study of the Environment, Santa Barbara, a nonprofit corporation that provides independent, science-based analyses of complex environmental issues. He has been a Professor of Biology and Director of the Program in Global Change at George Mason University, Fairfax, Virginia; Professor of Biology and Chairman of Environmental Studies at the University of California, Santa Barbara; and Professor of Systems Ecology at the Yale School of Forestry and Environmental Studies. His degrees are B.A. (Physics, University of Rochester); M.A. (Literature, University of Wisconsin); and Ph.D. (Biology, Rutgers University). He currently divides his time between New York City and California.

DATE DUE

FEB 2 2 2005			